AN ELEMENTARY PRIMER FOR GAUGE THEORY

An Elementary Primer For GAUGE THEORY

K. Moriyasu

Senior Research Physicist
University of Washington, Seattle

World Scientific Publishing Co Pte Ltd
P O Box 128
Farrer Road
Singapore 9128

ISBN 9971-950-83-9
9971-950-94-4 pbk

Printed in Singapore by Richard Clay (S. E. Asia) Pte. Ltd.

To My Parents

PREFACE

The understanding of nuclear and elementary particle physics has now reached a historical turning point. During the last decade, a revolution has quietly occurred – a revolution called "Gauge Theory". For the first time in 50 years, since the birth of modern nuclear physics, gauge theory allows us to understand how the fundamental forces of nature may be unified within a single coherent theory. The discovery of gauge theory rivals in importance the development of both relativity and quantum mechanics. In contrast to the situation less than 10 years ago, gauge theory now dominates nearly all phases of elementary particle physics today. Even the reasons for performing new experiments are now judged by their relevance for testing the predictions of gauge theory.

Clearly, such an exciting development should be widely accessible and understandable not only to theoreticians but also to experimental physicists, students and the "intelligent layman" as well. Like politics and war, gauge theory has become too important to be left only to the experts. Unfortunately, for the reader who wishes to first understand the basic physical ideas behind gauge theory, the published literature can present a daunting challenge. The reason for the difficulty is that gauge theory represents a totally new synthesis of quantum mechanics and symmetry ideas which have been applied to the entire field of elementary particle physics.

I believe that gauge theory can be appreciated by the non-expert; that is the raison d'etre for this primer. In order to emphasize the physics of gauge theory rather than the mathematical formalism, I have used a new intuitive approach and designed the text primarily for the reader with only a background in quantum mechanics. My goal in this primer is to hopefully leave the reader with an appreciation of the elegance and beauty of gauge theory.

This book was motivated by my own desire as a "non-expert" to learn something about gauge theory. Over a period of 4–5 years, I wrote a series of short pedagogical articles on gauge theory topics for the American and European Journals of Physics. These articles allowed me to test the ideas and the writing style for this primer. I also found that trying to satisfy the high standards of the referees for these journals encouraged me to develop much clearer explanations for many gauge theory topics. I am indebted to these referees who do their work in anonymity.

K. Moriyasu

Seattle,
July, 1983

CONTENTS

CHAPTER I

INTRODUCTION

> *. . . the best reason for believing in a renormalizable gauge theory of the weak and electromagnetic interactions is that it fits our preconceptions of what a fundamental field theory should be like.*
>
> S. Weinberg, 1974[1]

Modern gauge theory has emerged as one of the most significant and far-reaching developments of physics in this century. It has allowed us for the first time to realize at least a part of the age old dream of unifying the fundamental forces of nature. We now believe that electromagnetism, that most useful of all forces, has been successfully unified with the nuclear weak interaction, the force which is responsible for radioactive decay. What is most remarkable about this unification is that these two forces differ in strength by a factor of nearly 100 000. This brilliant accomplishment by the Weinberg-Salam gauge theory, and the insight gained from it, have encouraged the hope that all of the fundamental forces may be unified within a gauge theory framework. At the same time, it has been realized that the potential areas of application for gauge theory extend far beyond elementary particle physics. Although much of the impetus for gauge theory came from new discoveries in particle physics, the basic ideas behind gauge symmetry have also appeared in other areas as seemingly unrelated as condensed matter physics, non-linear wave phenomena and even pure mathematics. This diversity of interest in gauge theory indicates that it is in fact a very general area of study and not exclusively limited to elementary particles.

[1] S. Weinberg, *Rev. Mod. Phys.* **46**, 255 (1974).

In this primer for gauge theory, our purpose is to present an elementary introduction which will provide an adequate background for appreciating both the new theoretical developments and the experimental investigations into gauge theory. We have therefore adopted a very general pedagogical approach which should be useful for very different areas of physics. Like any new topic in physics, the study of gauge theory requires some familiarity with background material from other areas of physics and mathematics. Gauge theory represents a new synthesis of quantum mechanics and symmetry. At the same time, it is also a direct descendent of quantum electrodynamics; thus much of the published literature on modern gauge theory is written in the language of renormalizable quantum field theory which has proven so useful in electrodynamics.

In this primer, we have adopted the point of view that it is possible to learn the fundamentals of gauge theory by using a much simpler semiclassical approach. By semiclassical, we mean Maxwell's electromagnetism and old-fashioned Schrödinger quantum theory where the electromagnetic field is not second quantized. By using such an approach, we can emphasize the new physics of gauge theory without the added technical complexities of quantum field theory. A limitation of our approach is that we cannot discuss the problems associated with the quantization of gauge theory in any rigorous fashion. However, since these problems are among the most subtle and difficult in gauge theory, we feel that they can best be studied separately in more advanced treatments such as the excellent review of Abers and Lee.[2]

One essential requisite for the study of gauge theory is at least a nodding acquaintance with some of the terminology of group theory. The heart of any gauge theory is the gauge symmetry group and the crucial role that it plays in determining the dynamics of the theory. Fortunately, much of the necessary group theory is already familiar to physics students from the treatment of angular momentum operators in quantum mechanics. The essential difference in

[2]E. Abers and B. Lee, *Phys. Rep.* 9C, 2 (1973).

gauge theory is that the symmetry group is not associated with any physical coordinate transformation in space-time. Gauge theory is based on an "internal" symmetry. Therefore, one cannot speak of angular momentum operators, but must replace them with the more abstract concept of group generators. This is more than a mere change of labels because the generators have mathematical properties which were previously ignored in quantum mechanics but are very useful in gauge theory. In particular, we will see that the proper understanding of gauge invariance leads naturally to a geometrical description of gauge theory that is both highly intuitive and strongly resembles the familiar geometrical picture of general relativity. By exploiting this geometrical feature of gauge theory, we can often find much simpler interpretations of complicated physical phenomena such as gauge symmetry breaking, which is one of the most important ingredients of the Weinberg-Salam theory.

This primer is generally organized into three sections. The first section consisting of Chapters II through V introduces the concept of gauge invariance and describes the essential ingredients and physical assumptions which go into the building of a general gauge theory. We begin in Chapter II with the original inspiration of Hermann Weyl and briefly review why gauge theory was re-discovered three times in different physical contexts before the correct interpretation of gauge invariance was finally understood. In Chapter III, the geometrical interpretation of gauge symmetry is discussed and simple arguments are used to motivate and derive the essential mathematical building blocks of gauge theory. In Chapter IV, the familiar case of electromagnetism is used as a pedagogical guide for the construction of the Yang-Mills theory. The canonical Lagrangian formalism is introduced and the equations of motion are derived and discussed. The non-Abelian versions of Maxwell's equations are presented in Chapter V and compared with electromagnetism. The unique problems caused by non-linearity and lack of superposition in Yang-Mills gauge theories are discussed.

The second section of this primer deals with the new description of the electromagnetic, weak and strong forces as gauge theories.

We begin in chapter VI by briefly reviewing the salient experimental and theoretical features of the weak nuclear interaction which lead to the idea of a gauge theory. In chapter VII, we introduce the general formalism for understanding how gauge symmetry is broken by the Higgs mechanism. Several physical examples are discussed in detail to illustrate the dynamical mechanism responsible for symmetry breaking in different applications. In Chapter VIII, we present a simple introduction to the Weinberg-Salam theory of the unified weak and electromagnetic interactions. The procedure for unifying the weak and electromagnetic gauge symmetry groups is discussed in detail. Symmetry breaking of the weak interaction is introduced and the masses of the gauge vector are derived. In chapter IX, we present a brief introduction to the basic physical ideas in the color gauge theory of strong interactions. A simple intuitive argument is given for the new phenomena of "asymptotic freedom". By using an analogy between the vacuum of color gauge theory and a dielectric medium, it is shown how the so-called "running coupling constant" can be obtained.

The third section of this primer provides an introduction to some of the new "non-perturbative" features of gauge theory. In chapter X, we present a simple study of monopoles and vortices and explain how their properties can be understood as topological features of gauge theory.

In the appendix, we briefly summarize some of the key group theory terminology used in this primer.

CHAPTER II

THE REDISCOVERY OF GAUGE SYMMETRY

> . . *gauge invariance has no physical meaning, but must be satisfied for all observable quantities in order to ensure that the arbitrariness of* **A** *and* ϕ *does not affect the field strength.*
>
> Röhrlich, 1965[1]

2.1 Introduction

Gauge invariance was recognized only recently as the physical principle governing the fundamental forces between the elementary particles. Yet the idea of gauge invariance was first proposed by Hermann Weyl[2] in 1919 when the only known elementary particles were the electron and proton. It required nearly 50 years for gauge invariance to be "rediscovered" and its significance to be understood. The reason for this long hiatus was that Weyl's physical interpretation of gauge invariance was shown to be incorrect soon after he had proposed the theory. Gauge invariance only managed to survive because it was known to be a symmetry of Maxwell's equations and thus became a useful mathematical device for simplifying many calculations in electrodynamics. In view of the present success of gauge theory, we can say that gauge invariance was a classical case of a good idea which was discovered long before its time.

In this chapter, we present a brief historical introduction to the discovery and evolution of gauge theory. The early history of gauge theory can be divided naturally into old and new periods where the dividing line occurs in the 1950's. In the old period, we will return to Weyl's original gauge theory to gain insight into

[1]F. Röhrlich, *Classical Charged Particles* (Addison Wesley, Reading, Mass., 1965).
[2]H. Weyl, *Ann. Physik* **59**, 101 (1919).

several key questions. The most important question is what motivated Weyl to propose the idea of gauge invariance as a physical symmetry? How did he manage to express it in a mathematical form that has remained almost the same today although the physical interpretation has altered radically? And, how did the development of quantum mechanics lead Weyl himself to a rebirth of gauge theory?

The new period of gauge theory begins with the pioneering attempt of Yang and Mills[3] to extend gauge symmetry beyond the narrow limits of electromagnetism. Here we will review the radically new interpretation of gauge invariance required by the Yang-Mills theory and the reasons for the failure of the original theory. By comparing the new theory with that of Weyl, we can see that many of the original ideas of Weyl have been rediscovered and incorporated into the modern theory.

2.2 The Einstein Connection

In 1919, only two fundamental forces of nature were thought to exist – electromagnetism and gravitation. In that same year, a group of scientists also made the first experimental observation of starlight bending in the gravitational field of the sun during a total eclipse[4]. The brilliant confirmation of Einstein's General Theory of Relativity inspired Hermann Weyl to propose his own revolutionary idea of gauge invariance in 1919. To see how this came about, let us first briefly recall some basic ideas involved in relativity.

The fundamental concept underlying both special and general relativity is that there are no absolute frames of reference in the universe. The physical motion of any system must be described relative to some arbitrary coordinate frame specified by an observer, and the laws of physics must be independent of the choice of frame.

In special relativity, one usually defines convenient reference frames which are called "inertial", i.e. moving with uniform velocity.

[3] C. N. Yang and R. L. Mills, *Phys. Rev.* **96**, 191 (1954).
[4] H. von Kluber, *Vistas in Astronomy* **3**, 47 (1960).

For example, consider a particle which is moving with constant velocity v with respect to an observer. Let S be the rest frame of of the observer and S′ be an inertial frame which is moving at the same velocity as the particle. The observer can either state that the particle is moving with velocity v in S or that it is at rest in S′. The important point to be noted from this trivial example is that the inertial frame S′ can always be related by a simple Lorentz transformation to the the observer's frame S. The transformation depends only on relative velocity between particle and observer, not on their positions in space-time. The particle and observer can be infinitesimally close together or at opposite ends of the universe; the Lorentz transformation is still the same. Thus the Lorentz transformation, or rather the Lorentz symmetry group of special relativity, is an example of "global" symmetry.

In general relativity, the description of relative motion is much more complicated because one is dealing with the motion of a system in a gravitational field. For the sake of illustration, let us consider the following "gedanken" exercise for measuring the motion of a test particle which is moving through a gravitational field. The measurement is to be performed by a physicist in an elevator. The elevator cable has broken so that the elevator and physicist are falling freely[5]. As the particle moves through the field, the physicist determines its motion with respect to the elevator. Since both particle and elevator are falling in the same field, the physicist can describe the particle's motion as if there were no gravitational field. The acceleration of the elevator cancels out the acceleration of the particle due to gravity. This is a simple example of the principle of equivalence, which follows from the well-known fact that all bodies accelerate at the same rate in a given gravitational field (e.g. 9.8 m/sec^2 on the surface of the earth).

Let us now compare the physicist in the falling elevator with the observer in the inertial frame in special relativity. It might appear that the elevator corresponds to an accelerating or "non-inertial" frame that is analogous to the frame S′ in which the particle

[5]P. G. Bergmann, *Introduction to the Theory of Relativity* (Prentice-Hill, New York, 1946).

appeared to be at rest. However, this is not true because a real gravitational field does not produce the same acceleration at every point in space. As one moves infinitely far away from the source, the gravitational field will eventually vanish. Thus, the falling elevator can only be used to define a reference frame within an infinitesimally small region where the gravitational field can be considered to be uniform. Over a finite region, the variation of the field may be sufficiently large for the acceleration of the particle not to be completely cancelled.[a]

We see that an essential difference between special and general relativity is that a reference frame can only be defined "locally" or at a single point in a gravitational field. This creates a fundamental problem. To illustrate the difficulty, let us now suppose that there are many more physicists in nearby falling elevators. Each physicist makes an independent measurement so that the path of the particle in the gravitational field can be determined. How are the individual measurements to be related to each other? The measurements were made in separate elevators at different locations in the field. Clearly, one cannot perform an ordinary Lorentz transformation between the elevators. If the different elevators were related only by a Lorentz transformation, the acceleration would have to be independent of position and the gravitational field could not decrease with distance from the source.

Einstein solved the problem of relating nearby falling frames by defining a new mathematical relation known as a "connection". To understand the meaning of a connection, let us consider a 4-vector A_μ which represents some physically measured quantity. Now suppose that the physicist in the elevator located at x observes that A_μ changes by an amount $\mathrm{d}A_\mu$ and a second physicist in a different elevator at x' observes a change $\mathrm{d}A'_\mu$. How do we relate the changes $\mathrm{d}A_\mu$ and $\mathrm{d}A'_\nu$? In special relativity, the differential $\mathrm{d}A_\mu$ is also a vector like A_μ itself. Thus, the differential $\mathrm{d}A'_\nu$ in the

[a] A strongly varying gravitational field gives rise to "tidal" forces which can produce some unusual effects. For example, see the science fiction story *Neutron Star* by L. Niven (Ballantyne, New York, 1968).

the elevator at x' is given by the familiar relation

$$dA'_\nu = \frac{\partial x^\mu}{\partial x'^\nu} dA_\mu \qquad \text{(II - 1)}$$

where, according to the usual convention[b], the repeated index μ is summed over the values $= 0, 1, 2, 3$. The simple relation (II - 1) follows from the fact that the Lorentz transformation between x and x' is a linear transformation. What happens in general relativity? We can no longer assume that the transformation from x to x' is linear. Thus, we must write for dA'_ν the general expression

$$\begin{aligned} dA'_\nu &= \frac{\partial x^\mu}{\partial x'^\nu} dA_\mu + A_\mu\, d\left(\frac{\partial x^\mu}{\partial x'^\nu}\right) \\ &= \frac{\partial x^\mu}{\partial x'^\nu} dA_\mu + A_\mu \frac{\partial^2 x^\mu}{\partial x'^\nu \partial x'^\lambda} dx'^\lambda \quad . \end{aligned} \qquad \text{(II - 2)}$$

Clearly, the second derivatives $\partial^2 x^\mu / \partial x'^\nu \partial x'^\lambda$ will vanish if the x^μ are linear functions of the x'^ν.

How do we interpret the physical meaning of the extra term in (II - 2)? Such terms are actually quite familiar in physics. They occur in "curvilinear" coordinate systems. For example, suppose that two physicists are located on a circular path at the positions x, y and $x' = x + dx, y' = y + dy$ as shown in Fig. (2-1). The curved path could be the equator of the earth. Using the familiar curvilinear coordinates:

$$x = R\cos\phi , \qquad y = R\sin\phi \quad , \qquad \text{(II - 3)}$$

it can be easily seen that the differentials dx and dy depend on the coordinates x and y. Now suppose that the physicist at x, y measures

[b] The components of the 4-vector $A^\mu = (A^0, \mathbf{A})$ and $A_\mu = (A_0, \mathbf{A})$ with $A^0 = -A_0$. Vector components with upper and lower indices are related by $x_\mu = g_{\mu\nu} x^\nu$, where $g_{\mu\nu}$ is the metric tensor which appears in the definition of the invariant space-time interval $ds^2 = g_{\mu\nu}\, dx^\mu dx^\nu$. The components of $g_{\mu\nu}$ are $g_{11} = g_{22} = g_{33} = 1$, $g_{00} = -1$, and all other components are zero.

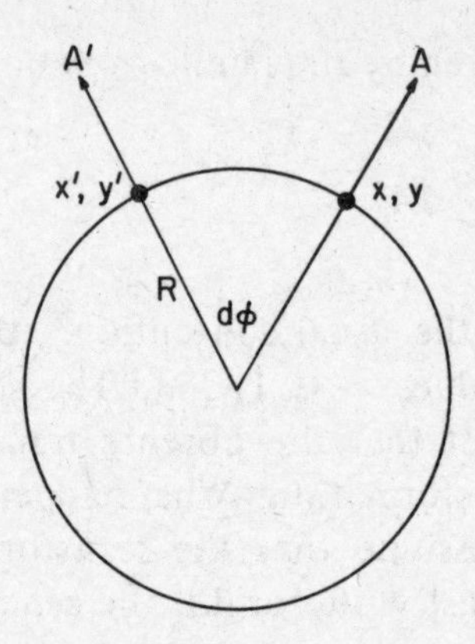

Fig. 2.1 A simple example of nonlinear effects in curvilinear coordinate system. Physicists at x, y and x', y', observe that vectors **A** and **A**$'$, which are perpendicular to the circle at their respective locations, do not point in the same direction.

a quantity which is described by a vector **A**. The vector **A** happens to be exactly perpendicular to the circular path at x, y as shown in Fig. (2.1). The direction of **A** could be interpreted as the "upward" direction when one is standing on the surface of the earth. However, the vector **A** is clearly different from the upward direction **A**$'$ at the position x', y' of the second physicist. The vector **A** at x, y is rotated by an angle $d\phi$ from the perpendicular direction **A**$'$ at x', y'. The difference between **A** and **A**$'$, assuming they have the same magnitude, is

$$|\mathbf{A}|d\phi = |\mathbf{A}|\frac{\partial\phi}{\partial x}dx + |\mathbf{A}|\frac{\partial\phi}{\partial y}dy \quad . \qquad \text{(II - 4)}$$

Let us compare this simple result with the extra term in (II - 2). We see that they both have the same general form, namely, they are proportional to the vector **A** itself and to the distance dx (and dy) between the points x and x'. This suggests that the second derivative in (II - 2) can be interpreted as a kind of coefficient which arises from "curvilinear" rather than linear transformations between x and x'. These curvilinear coefficients are denoted by the special symbol

$$\Gamma^{\mu}_{\nu\lambda} \equiv \frac{\partial^2 x^{\mu}}{\partial x'^{\nu}\,\partial x'^{\lambda}} \qquad \text{(II - 5)}$$

and are called the components of a "connection". They are also called affine connections or Christoffel symbols in texts on general relativity.[6]

Although we have used a simple angular coordinate system (II - 3) to interpret the meaning of the connection, it is important to note that the gravitational connection is not simply the result of using a curvilinear coordinate system. The value of the connection at each point in space-time is dependent on the properties of the gravitational field. The field determines the relative orientation of the different falling elevators in the same way that the "upward" direction on the surface of the earth varies from one position to another. The analogy with curvilinear coordinate systems merely indicates that the mathematical descriptions of free-falling frames and curvilinear coordinates are similar. It required the genius of Einstein to generalize this similarity and arrive at the revolutionary idea of replacing gravity by the curvature of space-time in general relativity.[7]

Let us briefly summarize the essential features of general relativity that Weyl would have utilized for his new gauge theory. First of all, general relativity involves a specific force, gravitation, which is not the case in special relativity. However, by studying the properties of coordinate frames just as in special relativity, one learns that only local coordinates can be defined in a gravitational field. This local property is required by the physical behavior of the field and leads naturally to the idea of a connection between local coordinate frames. Thus the essential difference between special and general relativity is that the former is a global theory while the latter is a local theory. This local property was the key to Weyl's gauge theory.

2.3 Weyl's Gauge Theory

Weyl went a step beyond general relativity and asked the

[6]C. W. Misner, K. S. Thorne and J. A. Wheeler, *Gravitation* (W. H. Freeman, San Francisco, 1973).
[7]Einstein, *The Meaning of Relativity* (Princeton University, New Jersey, 1955).

following question: if the effects of a gravitational field can be described by a connection which gives the relative orientation between local frames in space-time, can other forces of nature such as electromagnetism also be associated with similar connections? Generalizing the concept that all physical measurements are relative, Weyl proposed that the absolute magnitude or norm of a physical vector also should not be an absolute quantity but should depend on its location in space-time. A new connection would then be necessary in order to relate the lengths of vectors at different positions. This idea became known as scale or "gauge" invariance. It is important to note here that the true significance of Weyl's proposal lies in the local property of gauge symmetry and not in the particular choice of the norm or "gauge" as a physical variable. As we shall see, the assumption of locality is an enormously powerful condition that determines not only the general structure but many of the detailed features of gauge theory.

Weyl's gauge invariance can be easily expressed in mathematical form. We will use a notation which is somewhat more modern[8] than Weyl's original work. Let us consider a vector at position x with norm given by $f(x)$. If we shift the vector or transform the coordinates so that the vector is now at $x + \mathrm{d}x$, the norm becomes $f(x + \mathrm{d}x)$. Expanding to first order in $\mathrm{d}x$, we can write the norm as

$$f(x + \mathrm{d}x) = f(x) + \partial_\mu f \,\mathrm{d}x^\mu \quad , \qquad \text{(II - 6)}$$

where the abbreviation ∂_μ means $\partial/\partial x^\mu$. We now introduce a gauge change through a multiplicative scaling factor $S(x)$. This factor can be visualized as the change in size of a metre stick as shown in Fig. (2.2). The factor $S(x)$ is defined for convenience to equal unity at the position x. The scale factor at $x + \mathrm{d}x$ is then given by

$$S(x + \mathrm{d}x) = 1 + \partial_\mu S \,\mathrm{d}x^\mu \quad . \qquad \text{(II - 7)}$$

The norm of the vector at $x + \mathrm{d}x$ is then equal to the product of Eqs. (II - 6) and (II - 7). Keeping only first order terms in $\mathrm{d}x$, we obtain after a little algebra,

[8] C. N. Yang, *Physics Today*, **42** (June 1980).

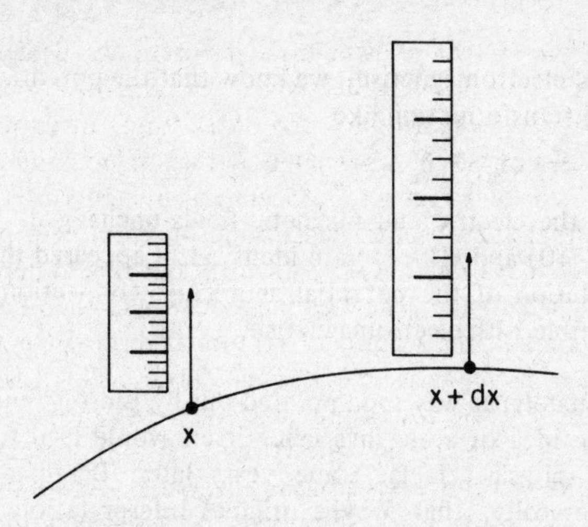

Fig. 2.2 The scale factor $S(x)$ in Weyl's gauge theory is illustrated by the change in length of metre stick from x to $x + \mathrm{d}x$.

$$Sf = f + (\partial_\mu S) f \,\mathrm{d}x^\mu + \partial_\mu f \,\mathrm{d}x^\mu \quad . \tag{II - 8}$$

For the special case of a constant vector, we see that the norm has changed by an amount

$$(\partial_\mu + \partial_\mu S) f \,\mathrm{d}x^\mu \quad . \tag{II - 9}$$

The derivative $\partial_\mu S$ is the new mathematical "connection" associated with the gauge change. Note that the connection $\partial_\mu S$ resembles the connection obtained from the simple example of curvilinear coordinates in (II - 4).

Weyl took the bold step of identifying the gauge connection $\partial_\mu S$ with the electromagnetic potential A_μ. One reason for this identification is that the connection itself transforms like a potential. It is straightforward to show that a second gauge change with a scale factor Λ will transform the connection as follows,

$$\partial_\mu S \longrightarrow \partial_\mu S + \partial_\mu \Lambda \quad . \tag{II - 10}$$

From classical electromagnetism, we know that the potential behaves under a gauge transformation like

$$A_\mu \longrightarrow A_\mu + \partial_\mu \Lambda \quad , \qquad \text{(II - 11)}$$

which leaves the electric and magnetic fields unchanged. Since the forms of (II - 10) and (II - 11) are identical, it appeared that Weyl's new interpretation of the potential as a gauge connection was perfectly compatible with electromagnetism.

Unfortunately, it was soon pointed out by Einstein and others[9] that the basic idea of scale invariance itself would lead to conflict with known physical facts. Some years later, Bergmann noted, somewhat ironically, that Weyl's original interpretation of gauge invariance would also be in conflict with quantum theory. The wave description of matter defines a natural scale for a particle through its Compton wavelength $\lambda = h/Mc$. Since the wavelength is determined by the particle's mass M, it cannot depend on position and thus contradicts Weyl's original assumption about scale invariance.

Despite the initial failure of Weyl's gauge theory, the idea of a local gauge symmetry survived. It was well known that Maxwell's equations were invariant under a gauge change. However, without an acceptable interpretation as some kind of physical coordinate transformation, gauge invariance was regarded as only an "accidental" symmetry of electromagnetism. The gauge transformation property in Eq. (II - 11) was interpreted as just a statement of the well known arbitrariness of the potential in classical physics. Only the electric and magnetic fields were considered to be real and observable. Gauge symmetry was retained largely because it was useful for calculations in both classical and quantum electrodynamics. As every student of physics knows, problems in electrodynamics can often be most easily solved by first choosing a suitable gauge, such as the Coulomb gauge or the Lorentz gauge, in order to make the equations more tractable.

[9]P. Bergmann, *Physics Today,* 44(March 1979).

2.4 Canonical Momentum and Electromagnetic Potential

A significant, but often overlooked, clue to the meaning of gauge invariance was provided by the formulation of electromagnetism as a classical Hamilton-Jacobi field theory. The most familiar consequence, well known to all physics students, is the simple recipe for replacing the momentum with the canonical momentum

$$p_\mu \longrightarrow p_\mu - eA_\mu \quad . \qquad \text{(II - 12)}$$

This replacement is all that is needed to build the electromagnetic interaction into the classical equations of motion.

The form of the canonical momentum is motivated by Hamilton's principle or the principle of least action. The essential idea is to obtain both Maxwell's equations and the equations of motion for charged particles from a single physical principle. This is accomplished formally by constructing a suitable Lagrangian density which contains all of the necessary information to describe the interaction of a charged particle with an electromagnetic potential A_μ. The derivation of the Lagrangian is found in many texts on electromagnetism[10] and gives

$$\mathcal{L} = \frac{1}{2}(p_\mu - eA_\mu)^2 - \frac{1}{4}F_{\mu\nu}F^{\mu\nu} \quad . \qquad \text{(II - 13)}$$

The first term in the Lagrangian, which explicitly contains the canonical momentum, is the kinetic energy. The second term, involving the Maxwell field tensor $F_{\mu\nu} = \partial_\mu A_\nu - \partial_\nu A_\mu$, gives the energy density of the electromagnetic field. The components of $F_{\mu\nu}$ are the electric and magnetic fields[c]:

$$F_{\mu\nu} = \begin{pmatrix} 0 & E_1 & E_2 & E_3 \\ -E_1 & 0 & -B_3 & B_2 \\ -E_2 & B_3 & 0 & -B_1 \\ -E_3 & -B_2 & B_1 & 0 \end{pmatrix} \quad . \qquad \text{(II - 14)}$$

[c] In some texts, the electric field components are multiplied by a factor of i.

[10] J. D. Jackson, *Classical Electrodynamics* (J. Wiley, New York, 1962).

Hamilton's principle is usually formulated in terms of a path that a dynamical system follows between two points. Given a system described by Lagrangian $L(q, t)$, where q is a generalized coordinate and t is time, the path integral defined by

$$S = \int L(q, t)\mathrm{d}t \tag{II - 15}$$

is called the "action". Hamilton's principle states that the path which the system must follow is the one for which the action S is a minimum. Although Hamilton's principle is purely classical in origin, the use of the path integral is now very familiar in quantum mechanics.

The minimum of the action S is found by varying the generalized coordinates q and velocities $\mathrm{d}q/\mathrm{d}t$ and setting the variation of S equal to zero[11]. The resulting equations are called the Euler-Lagrange equations of motion

$$\partial_\lambda \left[\frac{\partial \mathcal{L}}{\partial(\partial_\lambda q_\mu)} \right] - \frac{\partial \mathcal{L}}{\partial q_\mu} = 0 \quad , \tag{II - 16}$$

where the coordinate q_μ is now defined to be a four-vector. By using the Lagrangian (II - 13) and identifying the components of q_μ with spatial coordinates, (II - 16) yields the familiar Lorentz force law for charged particles. Alternatively, if we equate the q_μ with the electromagnetic potential A_μ, one obtains Maxwell's equations. Thus, within the Hamiltonian formalism, the electromagnetic potential acquires an added significance. The potential becomes an integral part of the canonical momentum, and it is thus treated as if it were a generalized coordinate in the Euler-Lagrange equations. Although these properties of the potential appear to be purely "formal", they played an essential role in the rediscovery of gauge invariance.

2.5 Quantum Mechanics and Gauge Theory

With the development of quantum mechanics, Hermann

[11] H. Goldstein, *Classical Mechanics* (Addison-Wesley, Reading, Mass, 1959).

Weyl[12] and others[13,14] realized that Weyl's original gauge theory could be given a new meaning. The essential clue was provided by the realization that the phase of a wavefunction could be a new local variable. Instead of a change of scale, a gauge transformation was re-interpreted as a change in the phase of the wavefunction,

$$\psi \longrightarrow \psi e^{-ie\lambda} \quad . \qquad \text{(II - 17)}$$

The familiar gauge transformation for the potential A_μ became

$$A_\mu \longrightarrow A_\mu - \partial_\mu \lambda \quad . \qquad \text{(II - 18)}$$

The Schrödinger equation for a charged particle in an electromagnetic field is left unchanged after the two transformations (II - 17) and (II - 18) are applied. The non-relativistic Schrödinger equation is written

$$\left[\frac{1}{2m}\left(-i\hbar\nabla - e\mathbf{A}\right)^2 + e\phi + V\right]\psi = i\hbar\frac{\partial\psi}{\partial t} \quad , \qquad \text{(II - 19)}$$

where the canonical momentum now appears as the quantum operator

$$-i\hbar\nabla - e\mathbf{A} \quad . \qquad \text{(II - 20)}$$

After the phase change (II - 17), there will be a new term proportional to $e\nabla\lambda$ from the operator $-i\hbar\nabla$ acting on the transformation wavefunction. This new term will be cancelled exactly by the gauge transformation of the potential according to (II - 18).

The phase of a wavefunction clearly satisfies the requirements for a new local variable. The previous objections to Weyl's original theory no longer apply because the phase is not directly involved in the measurement of a space-time quantity like the length of a vector. In the absence of an electromagnetic field, the amount of phase change can be assigned an arbitrary constant value since this would not affect any observable quantity. When an electromagnetic field

[12] H. Weyl, *Zeit. Physik* **56**, 330 (1929).
[13] V. Fock, *Zeit. Physik* **39**, 226 (1927).
[14] F. London, *Zeit. Physik* **42**, 375 (1927).

is present, a different choice of phase at each point in space can then be accommodated easily by interpreting the potential A_μ as a connection which relates phases at different points. The choice of a particular phase function $\lambda(x)$ will not affect any observable quantity provided that the gauge transformation for A_μ has just the right form (II - 18) so that the phase change and the change in potential cancel each other exactly. Thus, within the new environment of quantum mechanics, gauge invariance was re-discovered as invariance under a change in phase. The so-called "arbitrariness" formerly ascribed to the potential was now understood as the freedom to choose any value for the phase of a wavefunction without affecting the equations of motion.

2.6 The Aharonov-Bohm Effect

Nearly thirty years after Weyl's re-discovery of gauge invariance, Aharonov and Bohm[15] proposed a simple but ingenious experiment which directly challenged the idea that the potential A_μ could not produce any observable physical effects. A simple diagram of the experiment is shown in Fig. (2.3) from Feynman *et al*[16]. A beam of monoenergetic electrons from the source Q is diffracted by the

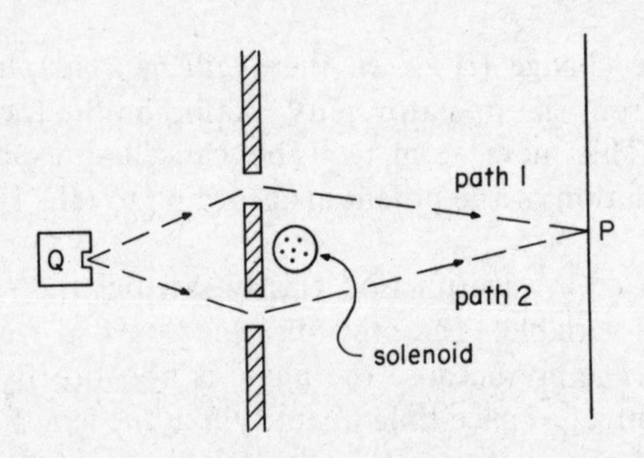

Fig. 2.3 Schematic diagram of the Aharonov-Bohm effect in a two-slit electron diffraction experiment. Electrons are produced by source Q, diffracted by the slits and their interference pattern is measured by detector at P. Solenoid is parallel to the slits and much smaller than spacing between slits.

[15] Y. Aharonov and D. Bohm, *Phys. Rev.* **115**, 485 (1959).

[16] R. P. Feynman, R. B. Leighton and M. Sands, *Feynman Lectures on Physics* Vol. **II** (Addison-Wesley, Reading, Mass., 1965).

two slits in the wall. The interference pattern produced by the two diffracted electron beams is measured at position P. Located behind the wall is a solenoid which is parallel to the slits. The diameter of the solenoid is much smaller than the distance between the slits so that the probability of the electrons passing through the solenoid is extremely small. Outside the solenoid, the magnetic field is exactly zero, but the potential is non-zero. When there is no current flowing in the solenoid, the normal two-slit interference pattern is seen at P. Aharonov and Bohm pointed out that after the current is turned on the vector potential alone should produce additional phase shifts in the wavefunctions of the electron beams. Thus the interference pattern observed at P should change, even though the magnetic field is zero outside the solenoid.

The predicted phase difference between the two electron beams is easily calculated. We will ignore the normal phase change produced by the two slits and determine only the additional change from the vector potential. Let ψ_0 be the electron wavefunction when there is no current in the solenoid. After the current is turned on, the Hamiltonian has the usual form

$$H = \frac{1}{2m}\left(-i\hbar\nabla - e\mathbf{A}\right)^2 \qquad \text{(II - 21)}$$

It was shown by Dirac[17] that the new wavefunction

$$\psi = \psi_0 \exp[-ieS/\hbar] \quad , \qquad \text{(II - 22)}$$

where

$$S = \int \mathbf{A}\cdot d\mathbf{x} \quad , \qquad \text{(II - 23)}$$

is the solution to the Schrödinger equation with the Hamiltonian in Eq. (II - 21). The integral in Eq. (II - 23) is evaluated along the paths 1 and 2 in Fig. (2.3). The new wavefunction ψ is just phase-shifted from ψ_0 by the quantity S which is the quantum analogue of the classical action from Hamilton's principle. At point P, the wavefunctions for the two electron beams are

[17] P. A. M. Dirac, *Proc. R. Soc.* **A133**, 60 (1931).

$$\psi_1 = \psi_0 \exp[-ieS_1/\hbar] \quad ,$$

$$\psi_2 = \psi_0 \exp[-ieS_2/\hbar] \quad . \qquad \text{(II - 24)}$$

The phase difference is easily seen to be

$$e(S_1 - S_2)/\hbar = \frac{e}{\hbar}\left[\int_1 \mathbf{A}\cdot d\mathbf{x} - \int_2 \mathbf{A}\cdot d\mathbf{x}\right] \quad . \qquad \text{(II - 25)}$$

By Stokes' theorem, this is directly proportional to the magnetic flux Φ in the solenoid.

Although Dirac's solution, Eq. (II - 23), leads to a correct description of the phase difference, it has the peculiar property that it is not single-valued. For example, if we move a test charge from Q to P along one path and then back to Q along a different path, the line integral in (II - 23) will give the net phase change at point Q. However, the value of the phase change will only be determined to within an arbitrary integer multiple of $2\pi e\Phi/\hbar$. These multiple phases actually have a physical interpretation. After leaving the point Q, if we move the test charge in a circular path around the solenoid N times before proceeding on to P, the phase difference will be changed by $2\pi e\Phi N/\hbar$. Thus each multiple phase represents a physically allowed path between Q and P. Even though it is highly unlikely that the electron will wrap N times around the solenoid, in quantum mechanics the contributions from these paths should be included in the path integral. A correct quantum mechanical treatment of these multiple phases was presented only recently. It has been shown by Berry[18] and others that the Aharonov-Bohm effect can be formulated as a standard scattering problem using a partial-wave expansion of multivalued plane waves. When the size of the solenoid is much larger than the de Broglie wavelength of the incident electrons, the calculation shows that the scattering amplitude is essentially dominated by the simple classical trajectories as expected.

An ingenious experimental test of the Aharonov-Bohm effect

[18] M. V. Berry, *Eur. Jour. Phys.* 1, 240 (1980).

was performed by Chambers[19] and his results confirmed the prediction of a phase shift. When the field strength within the solenoid was varied, the interference fringes shifted as expected even though there was no magnetic field in the regions where the electron beams passed. The amount of the phase shift also agreed quantitatively with the theoretical calculations. The Aharonov-Bohm effect provoked an extensive debate which we might find surprising today. The details of the controversy have been documented by Erlichson[20]. We quote Feynman *et al*[16]:

> *It is interesting that something like this can be around for thirty years but, because of certain prejudices of what is and is not significant, continues to be ignored.*

The Aharonov-Bohm effect clearly contradicted the accepted notion that only the electric and magnetic fields could produce observable effects. More important, it became evident that the potential had to be treated as a physical field that was also directly observable. The alternative would be to believe that the phase shift is produced by the magnetic field "acting at a distance" in direct conflict with relativity. Furthermore, the original objection to the potential, namely its arbitrariness, was no longer relevant. The Aharonov-Bohm effect showed that in quantum mechanics, it is the relative change in phase produced by the potential that is physically observable. Thus it became apparent that the vector potential field is not only observable but also much more "fundamental" than the electric and magnetic fields.

2.7 Electromagnetism as a Gauge Theory

Let us now summarize the status of gauge theory at the end of the old period. It is clear that the electromagnetic interaction of charged particles could be interpreted as a local gauge theory within the framework of quantum mechanics. In analogy with Weyl's original theory, the phase of a particle's wavefunction can be identified as a new physical degree of freedom which is dependent on the

[19] R. G. Chambers, *Phys. Rev. Lett.* **5**, 3 (1960).
[20] H. Erlichson, *Amer. Jour. Phys.* **38**, 162 (1970).

space-time position. The phase value can be changed arbitrarily by performing purely mathematical phase transformations on the wavefunction at each point. Therefore, as Weyl had argued originally, there must be some connection between phase values at nearby points. The role of this connection is played by the electromagnetic potential. This intimate relation between potential and the change in phase is clearly demonstrated by the Aharonov-Bohm effect. Thus by using the phase of a wavefunction as the local variable instead of the norm of a vector, electromagnetism can be interpreted as a local gauge theory very much as Weyl envisioned.

It is pertinent to ask why local gauge invariance was not recognised as a very general and fundamental principle of physics soon after its re-discovery? One apparent reason is that gauge invariance appeared to be a trivial or accidental symmetry of electromagnetism. Gauge transformations can be viewed as merely phase changes so that they look more like a property of quantum mechanics than electromagnetism. In addition, the symmetry defined by the gauge transformations does not appear to be "natural". The set of all gauge transformations forms a one-dimensional unitary group known as the U(1) group. This group does not arise from any form of coordinate transformation like the more familiar spin-rotation group SU(2) or the Lorentz group. Thus, one has lost the original interpretation proposed by Weyl of a new space-time symmetry.

The status of gauge theory was also influenced by the historical fact that Maxwell had formulated electromagnetism long before Weyl proposed the idea of gauge invariance. Therefore, unlike the general theory of relativity, the gauge symmetry group did not play any essential role in defining the dynamical content of electromagnetism. This sequence of events was to be completly reversed in the development of modern gauge theory.

2.8 Isotopic Spin and the New Gauge Theory

The basic concepts behind modern gauge theory originated from two important developments in the study of nuclear forces. First, the force between nucleons was known to have an extremely short range in contrast to electromagnetism. In 1935, Yukawa suggested that the nuclear force was mediated by the exchange of a new quantum in analogy with electrodynamics as shown in Fig. (2.4). However, unlike the photon which is massless, the new quantum was assumed to have a large mass which would explain the short range of the nuclear force. The mass of Yukawa's new quantum was introduced by replacing the familiar Coulomb potential

$$V = \frac{q}{4\pi\epsilon r} \qquad \text{(II - 26)}$$

with a new potential

$$V = \frac{g}{4\pi r} e^{-r/R} \quad , \qquad \text{(II - 27)}$$

where $R = \hbar/Mc$ is the range of the potential and M is the mass of the new quantum. In order to have a range of less than one fermi, the mass of the quantum would have to be greater than 200 MeV, which is not too far from the known mass 140 MeV, of the π-meson.

The second important discovery was the charge independence of the nuclear force[21]. The strength of the nuclear interaction between a proton and neutron, or between two protons or two neutrons, was measured and found to be the same. Charge independence was formulated as a new symmetry principle through the introduction of

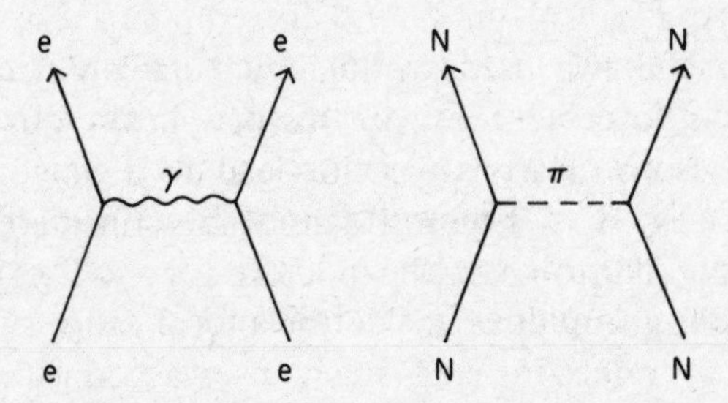

Fig. 2.4 Electromagnetic interaction between electron e is mediated by the exchange of a photon γ. The strong nuclear force between nucleons N is mediated by the exchange of mesons.

[21] J. M.Blatt and V. W. Weisskopf, *Theoretical Nuclear Physics* (J. Wiley, New York, 1952).

isotopic spin.

Heisenberg had suggested many years earlier that the proton and neutron be considered as the "up" and "down" states of an abstract isotopic spin in analogy with the ordinary spin states of the electron. This idea was extended to the nuclear force by adapting the familiar mathematical machinery associated with angular momentum. Charge independence was defined as the invariance of the nuclear force under a new SU(2) isotopic-spin rotation group. This led naturally to a conservation law for isotopic spin in analogy with the law for conservation of angular momentum. The isotopic spin group provided not only an economical way to classify the different charge states of elementary particles but also a fundamental principle for understanding the nuclear forces. These two complementary aspects of isotopic spin symmetry were best exemplified by the role of the π meson. When the charged and neutral π mesons were discovered, they were recognised as the three components of a state with isotopic spin one. The π-meson also was identified as the heavy "photon" proposed by Yukawa as the mediator of the strong nuclear interaction. By using the isotopic spin group to determine the coupling of the π-meson to the nucleons, one now could calculate nucleon scattering by exchanging a π-meson as shown in Fig. (2.4). Thus the basic concept that π-mesons were both eigenstates of isotopic spin and the carriers of the nuclear force became the keystone for the theory of the strong interactions.

Although Yukawa used an analogy between electromagnetism and the nuclear force to motivate his new heavy quantum, the exchange of π-mesons clearly does not lead to a gauge theory of the strong interaction. The π-meson cannot be considered as the true analogue of the photon for the nuclear force. The reason is that the isotopic spin group does not define a local gauge symmetry. The isotopic spin of π-mesons and nucleons was considered to be an internal quantum number independent of their space-time position. Isotopic spin was a "global" symmetry in the same sense as the Lorentz group of special relativity. Thus there was no need for a connection or an isotopic-spin potential field whose quantum could be identified with the π-meson. Various attempts were made[22, 23] to construct a gauge theory of π-meson exchange, but they were unsuccessful. Nevertheless, the discovery of isotopic spin symmetry provided the key ingredient for the next step in the development of gauge theory.

2.9 Yang-Mills Gauge Theory

In 1954, C. N. Yang and R. Mills took the bold step of proposing that the strong nuclear interaction be described by a field theory like electromagnetism which is exactly gauge invariant. They postulated that the local gauge group was the SU(2) isotopic-spin group. This idea was revolutionary because it changed the very concept of the "identity" of an elementary particle. If the nuclear interaction is a local gauge theory like electromagnetism, then there is a potential conflict with out intuitive notion of how to define a particle state. For example, let us assume that we can "turn off" the electromagnetic interaction so that we cannot distinguish the proton and neutron by electric charge. We also ignore the small mass difference. We must then label the proton as the "up" state of isotopic spin 1/2 and the neutron as the "down" state. But if isotopic spin invariance is an independent symmetry at each point in space-time, we cannot assume that the "up" state at one location is necessarily the same as the "up" state at any other point. The local isotopic spin symmetry allows us to choose arbitrarily which direction is "up" at each point without reference to any other point.

Given that the labelling of a proton or neutron is arbitrary at each point, once the choice has been made at one location, it is clear that some rule is then needed in order to make a comparison with the choice at any other position. The required rule, as Weyl proposed orginally, is supplied by a connection. A new isotopic-spin potential field was therefore postulated by Yang and Mills in analogy with the electromagnetic potential. However, the greater complexity of the SU(2) isotopic-spin group as compared to the

[22] A. Pais, *Physica* **19**, 869 (1953).
[23] B. Zumino, *Proc. 1969 CERN School of Physics* (CERN 69-29, Nov. 1969).

U(1) phase group means that the Yang-Mills potential will be quite different from the electromagnetic field. In electromagnetism, the potential provides a connection between the phase values of the wavefunction at different positions. In the Yang-Mills theory, the phase is replaced by a more complicated local variable that specifies the direction of the isotopic spin. In order to understand qualitatively how this leads to a connection, we need only to recall that the SU(2) isotopic-spin group is also the group of rotations in a 3-dimensional space[d]. As an example, let us visualize the "up" component of isotopic spin 1/2 as a vector in an abstract "isotopic spin space". An obvious way to relate the "up" states at different locations x and y is ask how much the "up" state at x needs to be rotated so that it is oriented in the same direction as the "up" state at y. This suggests that the connection between isotopic spin states at different points must act like an isotopic spin rotation itself. In other words, if a test particle in the "up" state at x is moved through the potential field to position y, its isotopic spin direction must be rotated by the field so that it is pointing in the "up" direction corresponding to the position y. We can immediately generalize this idea to states of arbitrary isotopic spin. Since the components of an isotopic spin state can be transformed into one another by elements of the SU(2) group, we can easily conclude that the connection must be capable of performing the same isotopic spin transformations as the SU(2) group itself. This novel idea that the isotopic spin connection, and therefore the potential, acts like the SU(2) symmetry group is the most important result of the Yang-Mills theory. This concept lies at the heart of local gauge theory. It shows explicitly how the gauge symmetry group is built into the dynamics of the interaction between particles and field.

How is it possible for a potential to generate a rotation in an internal symmetry space? To answer this question, we first must define the Yang-Mills potential more carefully in the language of the rotation group. A 3-dimensional rotation $R(\theta)$ of a wavefunction

[d]Technically, the SU(2) group is different from the group of 3-dimensional rotations, which is called O(3); the SU(2) group being the so-called "covering group" of O(3). For further details, see the discussion in appendix **A**.

is written

$$R(\theta)\psi = e^{-i\theta \mathbf{L}}\psi \quad , \tag{II - 28}$$

where θ is the angle of rotation and $\mathbf{L}$ is an angular momentum operator. Let us compare this rotation with the phase change of wavefunction after a gauge transformation. Clearly, the rotation has the same mathematical form as the phase factor of the wave-function. However, this does not mean that the potential itself is a rotation operator like $R(\theta)$. We noted earlier that the amount of phase change must also be proportional to the potential in order to ensure that the Schrödinger equation remains gauge invariant. To satisfy this condition, the potential must be proportional to the angular momentum operator $\mathbf{L}$ in (II - 28). Thus, the most general form of the Yang-Mills potential is a linear combination of the angular momentum operators

$$A_\mu = \sum_i A^i_\mu(x) L_i \quad , \tag{II - 29}$$

where the coefficients $A^i_\mu(x)$ depend on the space-time position. This relation indicates that the Yang-Mills potential is not a rotation, but rather is a "generator" of a rotation[e]. For the case of electro-magnetism, the angular momentum operator is replaced by a unit matrix and $A^i_\mu(x)$ is just proportional to the phase change $\partial_\mu \lambda$. The relation (II - 29) explicitly displays the dual role of the Yang-Mills potential as both a field in space-time and an operator in the isotopic-spin space.

We can immediately deduce some interesting properties of the Yang-Mills potential. For example, the potential must have three charge components corresponding to the three independent angular momentum operators L_+, L_- and L_3. The potential component which acts like a raising operator L_+ can transform a "down" state into an "up" state. We can associate this formal operation with real processes such as that shown in Fig. (2.5) where a neutron absorbs a unit of isotopic spin from the gauge field and turns into

[e]For further details of group generators, see the discussion in appendix A.

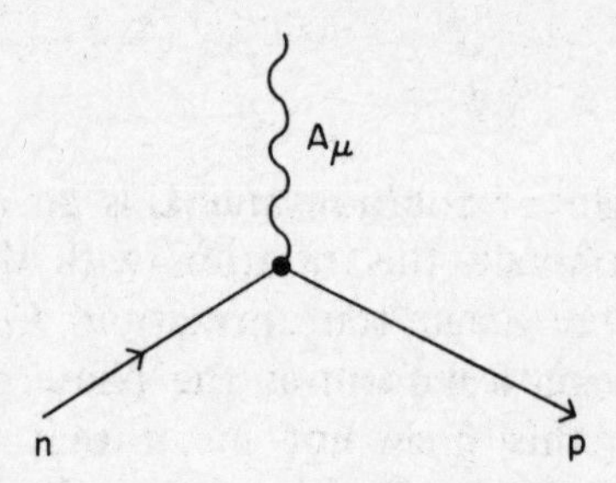

Fig. 2.5 A neutron n is transformed into a proton p by absorbing a unit of isotopic spin from the Yang-Mills gauge field A_μ. The field A_μ acts on the neutron state like an isotopic-spin "raising" operator.

a proton. This example indicates that the Yang-Mills gauge field must itself carry electric charge unlike the electromagnetic potential. The Yang-Mills field also differs in other respects from the electromagnetic field but they both have one property in common, namely, they have zero mass. The zero mass of the photon is well known from Maxwell's equations, but local gauge invariance requires that the mass of the gauge potential field be identically zero for any gauge theory. The technical reason is that the mass of the potential must be introduced into the Lagrangian through a term of the form

$$m^2 A_\mu A^\mu \quad . \tag{II - 30}$$

This guarantees that the correct equation of motion for a vector field will be obtained from the Euler-Lagrange equations. Unfortunately, the term given by (II - 30) is not invariant under a gauge transformation. The special transformation property of the potential will introduce extra terms in (II - 30) proportional to A_μ, which are not cancelled by the transformation of the wavefunction. Thus, the standard mass term is not allowed in the Yang-Mills Lagrangian. This means that the Yang-Mills gauge field must have exactly zero mass like the photon. The Yang-Mills field will therefore exhibit long-range behaviour like the Coulomb field and cannot reproduce the observed short range of the nuclear force. Since this conclusion appeared to be an inescapable consequence of local gauge invariance, the Yang-Mills theory was not considered to be an improvement on the already existing theories of the strong nuclear interaction.

Although the Yang-Mills theory failed in its original purpose, it established the foundation for modern gauge theory. The SU(2) isotopic-spin gauge transformation could not be regarded as a mere phase change; it required an entirely new interpretation of gauge invariance. Yang and Mills showed for the first time that local gauge symmetry was a powerful fundamental principle that could provide new insight into the newly discovered "internal" quantum numbers like isotopic spin. In the Yang-Mills theory, isotopic spin was not just a label for the charge states of particles, but it was crucially involved in determining the fundamental form of the interaction.

2.10 Gauge Theory and Geometry

The Yang-Mills theory revived the old ideas that elementary particles might have new degrees of freedom in some kind of "internal" space. By showing how these internal degrees of freedom could be unified in a non-trivial way with the dynamical motion in space-time, Yang and Mills discovered a new type of geometry in physics.

The new geometrical structure of gauge theory can be seen by comparing the Yang-Mills theory with our earlier discussion of general relativity. The essential role of the connection is evident in both gauge theory and relativity. Is there an analogue of the falling elevator or non-inertial coordinate frame in gauge theory? The answer is yes, but the local frame is located in an unfamiliar abstract space associated with the gauge symmetry group. To see how the gauge group defines an internal space, let us examine the examples of the U(1) phase group and the SU(2) isotopic spin group. For the U(1) group, the internal space consists of all possible values of the phase of the wavefunction. These phase values can be interpreted as angular coordinates in a 2-dimensional space. The internal symmetry space of U(1) thus looks like a ring as shown in Fig. (2.6), and the coordinate of each point in this space is just the phase value itself. The internal space defined by the SU(2) group is more complicated because it describes rotation in a 3-dimensional space.

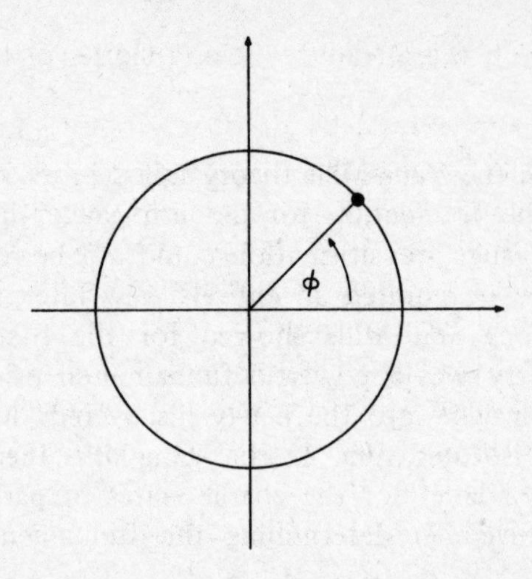

Fig. 2.6 The internal phase of the U(1) gauge group of electromagnetism. The phase ϕ defines an angular coordinate in the internal symmetry space.

We recall from our previous discussion, that all orientations of an isotopic spin state can be generated by starting from a fixed initial isotopic direction, which can be chosen as the isotopic spin "up" direction, and then rotating to the desired final direction. The values of the three angles which specify the rotation can be considered as the coordinates of a point inside an abstract 3-dimensional space. Each point corresponds to a distinct rotation so that the isotopic spin states themselves can be identified with the points in this angular space. Thus, the internal symmetry space of the SU(2) group looks like the interior of a 3-dimensional sphere.

The symmetry space of the gauge group provides the local non-inertial coordinate frame for the internal degrees of freedom. To an imaginary observer inside this internal space, the interaction between a particle and an external gauge field looks like a simple rotation of the local coordinates. The amount of the rotation is determined by the strength of the external potential, and the relative change in the the internal coordinates between two space-time points is just given by the connection as stated before. Thus, we see that there is a strong similarity between the geometrical description of relativity and this internal space picture of gauge theory.

The geometrical structure of gauge theory can be illustrated with a very convenient intuitive picture as shown in Fig. (2.7). Space-time is represented by the horizontal plane and the internal symmetry space is drawn vertically at each point. The vertical line in the figure depicts the case of a one-dimensional internal space like that of the U(1) group. The internal space is very appropriately called a "fiber" by mathematicians[24,25]. For the SU(2) group, we would have to attach a sphere at each space-time point. In this picture, the spatial location of a particle is given by a coordinate point in the horizontal plane and the orientation in the internal space in specified by angular coordinates in the "fiber" space. As the particle moves through space-time, it traces out a path in the internal space above the space-time trajectory. When there is no external gauge potential, the internal space path is completely arbitrary. When the particle interacts with an external gauge field, the path in the internal space is a continuous curve determined by the gauge potential.

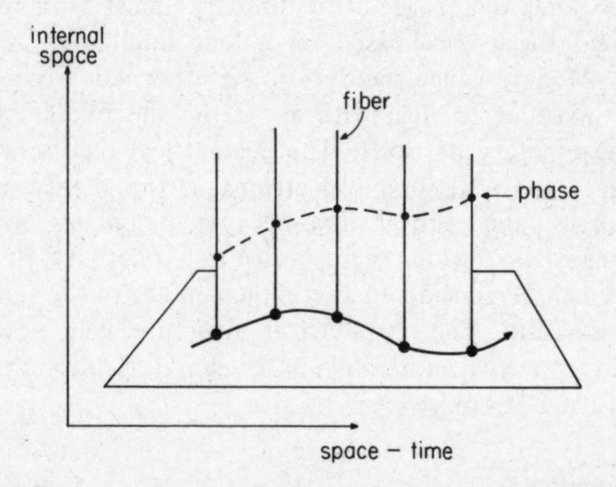

Fig. 2.7 Geometrical picture of the internal symmetry space. Space-time is the horizontal plane and the internal space direction is specified by phase angles in the internal space or "fiber".

[24] Y. Choquet-Bruhat, C. DeWitt-Morette and M. Dillard-Bleick, *Analysis, Manifolds and Physics*, revised edition (North-Holland, New York, 1982).
[25] W. Dreschsler and M. E. Mayer, *Fiber Bundle Techniques in Gauge Theories* (Springer-Verlag, Heidelberg, 1977).

The idea of using a gauge potential to "marry" together space-time with an internal symmetry space is a new concept in physics. Surprisingly, it turns out that the same type of geometrical structure had been invented by mathematicians at nearly the same time that gauge theory was re-discovered. In mathematical jargon, the new space formed by the union of 4-dimensional space-time with an internal space is called a "fiber bundle" space. The reason for this name is that the internal spaces or "fibers" at each space-time point can all be viewed as the same space because they can be transformed into each other by a gauge transformation. Hence, to a mathematician the total space is a collection or "bundle" of fibers.

Given that Yang and Mills developed their theory using the same terminology as electrodynamics, it is relevant to ask if there is any good reason to describe gauge theory in geometrical terms, other than to establish a historical link with relativity. The best reason for doing so is that the geometrical picture provides a valuable aid to the standard language of field theory. Most of the pedagogical aids in field theory are based on a long familiarity with electrodynamics. Modern gauge theory, on the other hand, requires a new approach in order to deal with all of the the fundamental forces between elementary particles. The geometrical picture can provide a common arena for discussing electromagnetism, the strong and weak nuclear forces, and even gravity, because it depends on only very general properties of gauge theory. Just as in relativity, the geometrical picture can be considered as complementary to the standard field theory formalism. The geometrical structure is a new facet of gauge theory that was not fully appreciated in electromagnetism but which has also been rediscovered.

2.11 Discussion

From a brief review of the early history of gauge theory, we have seen why it required nearly 50 years for gauge symmetry to be rediscovered. During that time, gauge invariance evolved from an accidental symmetry of Maxwell's equations to the basic physical principle governing the fundamental forces of nature. The original proposal of Weyl for a new space-time symmetry did not survive, but his ideas about new local degrees of freedom and a potential field which acts like a connection have become the foundations of modern gauge theory.

The development of gauge theory is a classic example of a good idea presented before its time. The correct interpretation of gauge theory was not possible without quantum mechanics. In electrodynamics, gauge invariance was regarded as only a convenient calculational device because it did not appear to be a genuine symmetry of nature. The relevance of the Aharonov-Bohm effect for gauge theory was unappreciated because it required that the potential be treated as a real physical field. And finally, the pioneering Yang-Mills gauge theory was regarded as a failure because exact gauge invariance was thought to be incompatible with a short-range force.

CHAPTER III

GAUGES, POTENTIALS AND ALL THAT

It still remains a mystery that the geometrical analysis which led to such a deep understanding of gravitation has no success elsewhere in physics.

F. Dyson, 1965[1]

3.1 Introduction

The essential building blocks of gauge theory are the gauge symmetry group, the gauge potential field which defines the connection, and the physical particles which are the sources of the gauge field and which also interact with each other via the gauge potential. In this chapter, we will see in detail how to formulate the gauge potential or connection for the case of a general non-Abelian gauge group like SU(2). Our approach will follow the geometrical point of view originally inspired by Weyl but adapted to the modern interpretation of gauge theory. The purpose in using this approach is to understand which features of a gauge theory can be determined by symmetry arguments alone independent of any application. In particular, we will see precisely how the concepts of a local symmetry and a geometrical connection determine the basic interactions of all particles in a gauge theory. Except for a little necessary group theory, our calculations will be performed by using simple geometrical arguments and making use of the time-honoured pedagogical device of a moving test charge to measure the effects of the external gauge field.

[1]F. Dyson, *Physics Today* (June, 1965).

3.2 Local Gauge Transformations

In the historical introduction to the Yang-Mills theory, we described qualitatively how the gauge group is associated with a connection. Any particle or system which is localized in a small volume and carries an internal quantum number like isotopic spin is considered to have a direction in the internal symmetry space. This internal direction can be arbitrarily chosen at each point in space-time. In order to compare these internal space directions at two different space-time points x and $x + \mathrm{d}x$, we need to define an appropriate connection which can tell us how much the internal direction at x differs from the direction at $x + \mathrm{d}x$. This connection must be capable of relating all possible directions in the internal space to each other. The most obvious way to relate two directions is to find out how much one direction has to be rotated so that it agrees with the other direction. The set of all such rotations forms a symmetry group; thus, the connection between internal space directions at different points must act like a symmetry group as well.

Our problem now is to see how a symmetry group transformation can lead us to a connection which we will be able to identify with a gauge potential field. We will begin by writing the general form of a local symmetry transformation for an arbitrary (non-Abelian) group,

$$U\Psi = \exp\left(-\mathrm{i}q \sum_k \theta^k(x) F_k\right)\Psi \quad . \qquad \text{(III - 1)}$$

The "local" nature of the transformation is indicated by the parameters $\theta^k(x)$ which are continuous functions of x. The constant q is the electric charge for the U(1) gauge group or a general "coupling constant" for an arbitrary gauge group. This is the only way in which the charge enters directly into the calculation. The general transformation (III - 1) is identical to the usual form of an ordinary spatial rotation if we identify the position-dependent parameter $\theta^k(x)$ with rotation angles. The F_k are the generators of the internal symmetry group and satisfy the usual commutation relations

$$[F_i, F_j] = \mathrm{i}c_{ijk}F_k \quad , \qquad \text{(III - 2)}$$

where the structure constants c_{ijk} depend on the particular group. For the isotopic-spin rotation group SU(2), the generators F_k are just the familiar angular momentum operators. To see how the transformation (III - 1) defines a connection, let us consider the following simple gedanken operation. We will take a test particle described by a wavefunction $\psi(x)$ and move it between two points x and $x + dx$ in space-time, and analyze how its direction changes in the internal symmetry space. The internal direction at x is initially chosen to have the angles $\theta^k(x)$. As the test particle moves away from x, the internal direction changes in some continuous way until it reaches $x + dx$ where it has a new internal direction given by angles $\theta^k(x + dx)$ as shown in Fig. (3.1). For an infinitesimal distance dx, this change can be described by the effect of the transformation (III - 1) action on $\psi(x)$ and producing a rotation of the internal direction equal to the difference $d\theta^k = \theta^k(x + dx) - \theta^k(x)$. This rotation clearly gives us just what we need, namely, a connection between internal space directions at different points in space-time. We also see that this connection will involve the derivative of a quantity just like the connection defined by Weyl. In this case, the quantities are the internal rotation angles $\theta^k(x)$. Thus, we see this is a straightforward generalization of the phase of a wavefunction to a set of angles which specify the internal direction.

3.3 Connections and Potentials

Now let us see explicitly how to calculate the connection from the symmetry transformation (III - 1) by moving the charged test

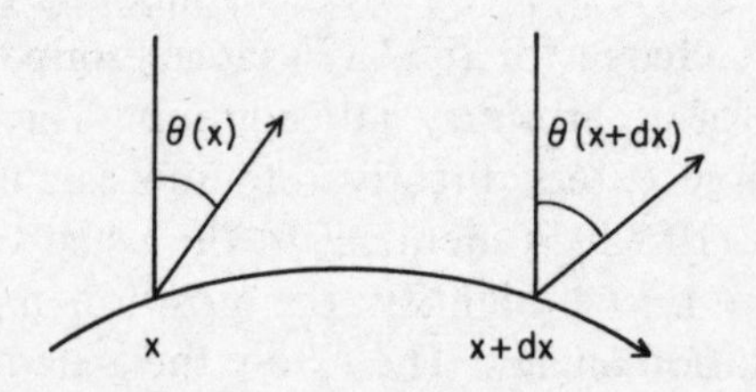

Fig. 3.1 Rotation of the internal space direction of a moving test particle. The internal angles change from $\theta(x)$ to $\theta(x + dx)$ as the test charge moves from x to $x + dx$.

particle through an external potential field. For pedagogical purposes, we will explicitly separate the particle's wavefunction $\psi(x)$ into external and internal parts. Let us write

$$\psi(x) = \sum_{\alpha} \psi_{\alpha}(x) u_{\alpha} \quad , \tag{III - 3}$$

where the u_α form a set of a "basis vectors" in the internal space. The index α is an internal label such as the components of isotopic spin. The basis u_α is analogous to the local non-inertial frame in relativity. The external part $\psi_\alpha(x)$ is then a "component" of $\psi(x)$ in the basis u_α. Under an internal symmetry transformation, they transform in the usual way

$$\psi_{\beta} = U_{\beta\alpha} \psi_{\alpha} \quad , \tag{III - 4}$$

where $U_{\beta\alpha}$ is the some matrix representation of the symmetry group. We assume, though it is not necessary, that the representation is "irreducible" so that the particle has a unique charge or isotopic spin. The decomposition in Eq. (III - 3) is particularly useful because it will allow us to interpret the effect of the external potential field on the particle as a precession of the internal basis.

Now, when the test particle moves from x to $x + dx$ through the external potential field, $\psi(x)$ changes by an amount $d\psi$ given by

$$d\psi = \psi(x + dx) - \psi(x) \quad . \tag{III - 5}$$

In general, $d\psi$ must contain both the change in the external x-dependent part of $\psi_\alpha(x)$ and the change in the internal space basis u_α. From Eq. (III - 3) we can expand $d\psi$ to first order in dx as

$$d\psi = \sum_{\alpha} \left[(\partial_{\mu} \psi_{\alpha}) dx^{\mu} u_{\alpha} + \psi_{\alpha} du_{\alpha} \right] \quad . \tag{III - 6}$$

The second term contains the change du_α in the internal space basis. This term is given by the connection which we discussed above; it describes the effect of the external potential field on the internal space direction of the particle.

We now need to calculated the change du_α in the internal space basis. We stated earlier that the connection between the internal space direction at different space-time points is just given by an internal rotation. In this case, the internal directions are specified by a set of basis vectors, so we must calculate the change du_α from an infinitesimal internal rotation $U(dx)$ which is associated with the external displacement dx.

From Eq. (III - 1), we calculate the infinitesimal internal rotation $U(dx)$,

$$U(dx) = \exp\left[-iq \sum_k d\theta^k F_k\right] \quad , \qquad \text{(III - 7a)}$$

$$d\theta^k = (\partial_\mu \theta^k) dx^\mu \quad , \qquad \text{(III - 7b)}$$

which rotates the internal basis u by an amount du,

$$U(dx)u = u + du \quad . \qquad \text{(III - 8)}$$

The generators F_k act like matrix operators on the column basis vector u_α so we can write

$$U(dx)u_\alpha = \exp\left[-iq \sum_k (\partial_\mu \theta^k) dx^\mu (F_k)_{\alpha\beta}\right] u_\beta \ . \qquad \text{(III - 9)}$$

Expanding $U(dx)$ to first order in dx, we obtain

$$u_\alpha + du_\alpha = \left[\delta_{\alpha\beta} - iq \sum_k (\partial_\mu \theta^k) dx^\mu (F_k)_{\alpha\beta}\right] u_\beta \quad , \qquad \text{(III - 10)}$$

which then gives for the change in the basis,

$$du_\alpha = -iq \sum_k (\partial_\mu \theta^k) dx^\mu (F_k)_{\alpha\beta} u_\beta \quad . \qquad \text{(III - 11)}$$

The net change du_α will give us the connection that we have been seeking. Let us therefore introduce the new "connection operator

$$(A_\mu)_{\alpha\beta} = \sum_k (\partial_\mu \theta^k)(F_k)_{\alpha\beta} \quad . \qquad \text{(III - 12)}$$

We thus finally obtain for the total change $d\psi$,

$$d\psi = \sum_{\alpha\beta} \left[(\partial_\mu \psi_\alpha)\delta_{\alpha\beta} - iq(A_\mu)_{\alpha\beta}\psi_\alpha \right] dx^\mu u_\beta \quad , \qquad \text{(III - 13)}$$

where we have put in $\delta_{\alpha\beta}$ in order to factor out the basis vector u_β. Now, we can factor the change $d\psi$ into its own external and internal parts

$$\begin{aligned} d\psi &= \sum_\beta (d\psi)_\beta u_\beta \\ &\equiv \sum_\beta (D_\mu \psi_\beta) dx^\mu u_\beta \quad . \end{aligned} \qquad \text{(III - 14)}$$

The new operator D_μ is a generalized form of derivative known as the gauge covariant derivative which describes the changes in both the external and internal parts of $\psi(x)$. Thus we get for $D_\mu \psi_\beta$ from Eq. (III - 13)

$$D_\mu \psi_\beta = \sum_\alpha \left[\delta_{\beta\alpha}\partial_\mu - iq(A_\mu)_{\beta\alpha} \right] \psi_\alpha \quad . \qquad \text{(III - 15)}$$

For the case of the electromagnetic gauge group U(1), the internal space is one-dimensional so that Eq. (III - 15) reduces to

$$D_\mu \psi = (\partial_\mu - iqA_\mu)\psi \quad . \qquad \text{(III - 16)}$$

This is just the so-called "canonical momentum" which is familiar from ordinary electromagnetism. What may not be so familiar is the geometrical derivation we have used and the interpretation of the canonical momentum as the gauge covariant derivative. We can also deduce easily from the example of the U(1) group that the connection operator defined in Eq. (III - 12) should be identified as the generalized version of the vector potential field A_μ. Thus, we conclude that the external potential field is indeed a connection in the internal symmetry space.

3.4 The Vector Potential Field

In order to complete the identification of the connection operator with the usual vector potential, we can show that it has the correct transformation properties under gauge transformations for fixed position x. A simple way to do this is by seeing how $D_\mu \psi$ transforms. Since $D_\mu \psi$ is just the "total" derivative of ψ (external and internal), we clearly expect $D_\mu \psi$ to transform in the same way as ψ itself. Hence, we have

$$\psi' = U\psi \quad , \tag{III - 17a}$$

$$D'_\mu \psi' = U(D_\mu \psi) \quad . \tag{III - 17b}$$

The second equation becomes

$$(\partial_\mu - iqA'_\mu)U\psi = U(\partial_\mu - iqA_\mu)\psi \tag{III - 18}$$

which in turn can be solved for A'_μ

$$A'_\mu = UA_\mu U^{-1} - \frac{i}{q}(\partial_\mu U)U^{-1} \quad . \tag{III - 19}$$

Again, let us consider the example of electromagnetism. The gauge transformation U can be written

$$U = e^{-iq\lambda(x)} \tag{III - 20}$$

for fixed x, and we see that

$$A'_\mu = A_\mu - \partial_\mu \lambda \quad , \tag{III - 21}$$

which confirms that $(A)_{\alpha\beta}$ is the generalized form of the usual vector potential.

It is important to note that the fixed-x gauge transformation, Eq. (III - 20), is just an ordinary internal symmetry transformation like an old-fashioned isotopic spin rotation. It is analogous to a special relativity transformation in the freely falling elevator. This suggests a simple geometrical interpretation for the transformation of A_μ. The gauge transformation rotates the internal space coordinates. Since A_μ is an external field, the transformed field A_μ can be regarded as the field seen by an internal observer in a coordinate frame which has been rotated. Thus, the so-called arbitrariness in the vector potential A_μ can be understood as an arbitrariness in the absolute orientation of the internal coordinate frame. This is also clear from the fact that a connection only specifies "relative" orientations of coordinates, not absolute values.

Let us now summarize a few interesting properties of the vector potential. From our derivation, it is clear that the potential is both an external field and an internal space operator. From the definition, Eq. (III - 12), the external part of A_μ is the four-vector field $\partial_\mu \theta(x)$. Thus the vector potential looks like a spin-one field independent of the particular choice of internal symmetry group. This explains, though not completely rigorously, why the photon and all other gauge fields have spin one (except gravitation).

For non-Abelian gauge groups like SU(2), the vector potential field can carry internal "charge" unlike the case of the photon. This follows again from the definition, Eq. (III - 12), which shows that the internal operator part of A_μ is a linear combination of group generators F_k. For example, the vector potential for the isotopic spin group would have three components which could be chosen as $I_\pm$ and I_3. The component I_+ would act like an isotopic spin raising operator which could transform a neutron into a proton and therefore would have to carry one unit of "isotopic charge". In general, the number of components of the internal part of the vector potential field is equal to the dimension of the symmetry group.

3.5 Choosing a Gauge

Let us now see what the preceding geometrical formalism can tell us about the physics concepts involved in the choice of a gauge such as the Lorentz or Coulomb gauge. We know that the usual reason for selecting a particular gauge is to simplify a calculation or to explicitly display an interesting feature of a problem. This freedom to choose a gauge is often cited as just another example of the arbitrariness in the vector potential.

The choice of a gauge actually involves both gauge invariance and Lorentz invariance simultaneously. A particular gauge usually imposes a constraint on the vector potential, such as $\nabla \cdot \mathbf{A} = 0$ for the Coulomb gauge. In general, such equations are obviously not

Lorentz invariant. Yet we know that two observers, each in a different inertial frame, can both choose the Coulomb gauge for the same electromagnetism problem. The electric and magnetic fields observed in the two different frames can then be related by the usual Lorentz transformations between two frames. This example points out the important fact that the space-time location x at which the internal coordinate $\theta(x)$ is evaluated is also not a fixed position. A Lorentz transformation which changes the spatial coordinate in the inertial frame also affects the value of the internal angles and can be interpreted as an internal rotation. Thus, regardless of whether the coordinate change is associated with a Lorentz transformation of observers or an actual movement of the particle in the external field, the effect on the internal space is the same; it is rotated by a gauge transformation.

A Lorentz transformation between two inertial frames is therefore always associated with a gauge transformation as well. Thus, the vector potentials observed in the two frames are related by

$$A'_\mu = L_\mu{}^\nu A_\nu - \partial_\mu \lambda \quad , \tag{III - 22}$$

where $L_\mu{}^\nu$ is the Lorentz transformation. This shows that the vector potential actually does not transform like an ordinary four-vector under a Lorentz transformation. It picks up an extra term $\partial_\mu \lambda$ due to the rotation in the internal space. This interesting fact is well known in quantum field theory[2] but it is rarely mentioned in ordinary electromagnetism.

We can now see exactly what is involved for a particular choice of gauge. In the Lorentz gauge, $\partial^\mu A_\mu$ is required to be invariant,

$$\partial'^\mu A'_\mu = \partial^\mu A_\mu \quad , \tag{III - 23}$$

even though A_μ is not a true four-vector. From Eq. (III - 22), we see that this is possible only if

$$\partial^\mu \partial_\mu \lambda = 0 \quad , \tag{III - 24}$$

[2] J. D. Bjorken and S. Drell, *Relativistic Quantum Fields,* 73 (McGraw-Hill, New York, 1965).

or, equivalently,

$$\nabla^2 \lambda - \frac{\partial^2 \lambda}{\partial t^2} = 0 \quad , \tag{III - 25}$$

which is the familiar equation for λ in the Lorentz gauge. Thus we see that the choice of the Lorentz gauge is a requirement that the effect of the internal space precession be eliminated so that $\partial^\mu A_\mu$ can be treated as if it were a relativistic invariant. By the same reasoning, other gauges like the Coulomb gauge are not invariant because the gauge condition does not completely eliminate the extra internal precession term. Thus in the Coulomb gauge,

$$\nabla' \cdot \mathbf{A}' = \nabla \cdot \mathbf{A} = 0 \quad , \tag{III - 26}$$

but $\mathbf{A}$ and $\mathbf{A}'$ are not related by a simple Lorentz transformation. An additional gauge rotation is required. An explicit example of such a gauge rotation has been calculated by France.[3]

3.6 Maxwell's Field Tensor and Stoke's Theorem

We know from classical electromagnetism that the flux through the area bounded by a closed path is given by Stokes' theorem

$$\text{Flux} = \oint \mathbf{A} \cdot d\mathbf{x} \quad , \tag{III - 27}$$

where the line integral is taken around the path. From our preceding discussion, it is straightforward to see that the line integral over the potential A_μ can be interpreted geometrically as the net change in the internal direction of a test particle which has been moved around the closed path. The right-hand side of Stokes' theorem is therefore an expression for the phase change of the particle's wavefunction. On the other hand, the flux is determined by the value of the Maxwell field tensor $F_{\mu\nu}$ on the surface enclosed by the path. Does this mean that the Maxwell tensor $F_{\mu\nu}$ can be derived geometrically from Stokes' theorem by moving a test charge around a closed path through the external gauge field?

We can answer the above question by performing the following

[3]P. W. France, *Am. Jour. Phys.* **44**, 798 (1976).

simple exercise. We will move a test particle through successive displacements dx and dy around the sides of a closed path and calculate the net change in the internal direction. For simplicity, we can take dx and dy to be the sides of a parallelogram as shown in Fig. (3.2). As the test particle is displaced along each of the sides, it is acted upon by an infinitesimal gauge transformation of the form

$$U(dx) = 1 - iqA_\mu(x)dx^\mu \quad . \tag{III - 28}$$

In order to calculate the net phase change in the test particle's wavefunction, we must apply the transformation (III - 28) four successive times around the parallelogram. However, this does not mean that we can simply add up the phase changes from the four gauge transformations. The gauge transformation for each side acts on the wavefunction at different positions. This means that the phase change along each side is referred to a different initial phase value. In order to add up the phase changes around the closed path, we must be careful to avoid any spurious contributions from mismatched phases between the sides. Thus, the gauge transformations along each side must all be expressed in terms of the coordinates at a single common point x.

The gauge transformation for the displacement $x \to x + dx$ along side 1 is given by

$$U_x(dx) = [1 - iqA_\mu(x)dx^\mu] \quad . \tag{III - 29}$$

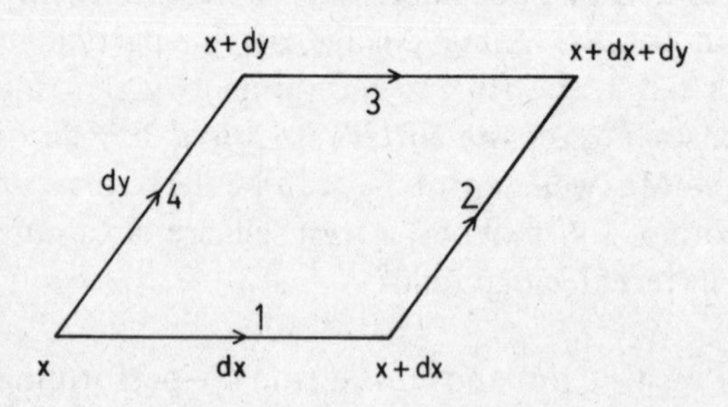

Fig. 3.2 Parallelogram used to define paths in a gauge field for the derivation of Maxwell field tensor. The test charge is moved from x to $x + dx + dy$ along paths 1 and 2.

For side 2, the corresponding gauge transformation for the displacement $x + \mathrm{d}x \rightarrow x + \mathrm{d}x + \mathrm{d}y$ is

$$U_{x+\mathrm{d}x}(\mathrm{d}y) = [1 - \mathrm{i}qA_\nu(x + \mathrm{d}x)\mathrm{d}y^\nu] \quad . \tag{III - 30}$$

To take into account the phase difference between sides 1 and 2, we must write $A(x + \mathrm{d}x)$ in terms of the coordinates at x. Expanding $A_\nu(x + \mathrm{d}x)$ to first order in $\mathrm{d}x$, we obtain

$$U_{x+\mathrm{d}x}(\mathrm{d}y) = [1 - \mathrm{i}qA_\nu(x)\mathrm{d}y^\nu - \mathrm{i}q\partial_\mu A_\nu(x)\mathrm{d}x^\mu \mathrm{d}y^\nu] \quad . \tag{III - 31}$$

We are now allowed to combine the transformations for sides 1 and 2 by taking the product

$$\begin{aligned} U_{x+\mathrm{d}x}(\mathrm{d}y)\, U_x(\mathrm{d}x) = {} & 1 - \mathrm{i}qA_\mu(x)\mathrm{d}x^\mu - \mathrm{i}qA_\nu(x)\mathrm{d}y^\nu \\ & - q^2 A_\nu(x)A_\mu(x)\mathrm{d}y^\nu \mathrm{d}x^\mu \\ & - \mathrm{i}q\partial_\mu A_\nu(x)\mathrm{d}y^\nu \mathrm{d}x^\mu \quad , \end{aligned} \tag{III - 32}$$

where we have kept terms only up to order $\mathrm{d}x\,\mathrm{d}y$. The gauge transformations for sides 3 and 4, which correspond to the displacements $x + \mathrm{d}x + \mathrm{d}y \rightarrow x + \mathrm{d}y$ and $x + \mathrm{d}y \rightarrow x$, are calculated by continuing in the same way. However, we can save a good deal of tedious algebra in the calculation for sides 3 and 4 by starting again at point x and going in the opposite order. The benefit of this procedure is that we can simply use the result (III - 32) by changing some labels. The net phase change can then be obtained by subtracting sides 3 and 4 from sides 1 and 2. We can write down the gauge transformation for sides 3 and 4 as

$$\begin{aligned} U_{x+\mathrm{d}y}(\mathrm{d}x)\, U_x(\mathrm{d}y) = {} & 1 - \mathrm{i}qA_\nu(x)\mathrm{d}y^\nu - \mathrm{i}qA_\mu(x)\mathrm{d}x^\mu \\ & - q^2 A_\mu(x)A_\nu(x)\mathrm{d}x^\mu \mathrm{d}y^\nu \\ & - \mathrm{i}q\partial_\nu A_\mu(x)\mathrm{d}x^\mu \mathrm{d}y^\nu \quad . \end{aligned} \tag{III - 33}$$

When we compare (III - 32) with (III - 33) we note that the $\mathrm{d}x^\mu$ and $\mathrm{d}y^\nu$ commute but the potential field components $A_\mu(x)$ and

and $A_\nu(x)$ do not because they are different combinations of the group generators F_k. In addition, the derivatives $\partial_\mu A_\nu(x)$ and $\partial_\nu A_\mu(x)$ are not equal in general. Thus, as we might have expected, the gauge transformation (III - 32) for sides 1 and 2 does not produce the same change in phase as the transformation (III - 33) for sides 3 and 4.

We can calculate the net change in the internal direction from the successive gauge transformations along the four sides of the parallelogram. This is equal to the difference between (III - 32) and (III - 33),

$$U(\mathrm{d}y)U(\mathrm{d}x) - U(\mathrm{d}x)U(\mathrm{d}y)$$

$$= -\mathrm{i}q\left\{\partial_\mu A_\nu - \partial_\nu A_\mu - \mathrm{i}q[A_\mu, A_\nu]\right\}\mathrm{d}x^\mu \mathrm{d}y^\nu \quad . \qquad \text{(III - 34)}$$

We now compare this result with Stokes' theorem. Since $\mathrm{d}x^\mu \mathrm{d}y^\nu$ is the surface area of the parallelogram, we identify the non-Abelian version of the Maxwell field tensor as the new operator

$$F_{\mu\nu} = \partial_\mu A_\nu - \partial_\nu A_\mu - \mathrm{i}q[A_\mu, A_\nu] \quad . \qquad \text{(III - 35)}$$

The extra term involving the commutator of the A_μ components arises for non-Abelian groups like SU(2) because the group generators do not commute. For the U(1) gauge group electromagnetism, the commutator vanishes and $F_{\mu\nu}$ reduces to its familiar form.

We see that the net result from our manipulation of the test particle is not only derivation of the Maxwell field tensor but also a geometrical interpretation of Stokes' theorem. In classical electromagnetism, Stokes' theorem is a useful mathematical relation between a vector field and its curl. In gauge theory, we see that Stokes' theorem is also an equation for the net change in the internal direction of a particle around a closed path. Another consequence of our calculation is that it explicitly demonstrates how the change in the internal direction depends on the particular path which is taken. We saw that the infinitesimal gauge transformation for sides 1 and 2 of the parallelogram was not the same as the transformation for sides 3 and 4. Thus, if we move the test particle along different paths between the same end points, the change in the internal directions will be different. We recall that this result was the reason for the shift of the interference pattern in the Aharonov-Bohm effect. The preceding calculation shows that the Aharonov-Bohm effect should also appear in general non-Abelian gauge theories.[4]

[4] T. T. Wu and C. N. Yang, *Phys. Rev.* D**12**, 3845 (1975).

CHAPTER IV

YANG-MILLS GAUGE THEORIES

> *. . . it is difficult to formulate the principle of minimal electromagnetic coupling in a mathematically unambigiuous way . . . . . It appears that the electromagnetic coupling generated by* $\partial_\mu \to \partial_\mu - ieA_\mu$ *is minimal when the free Lagrangian is chosen to be "minimal".*
>
> Sakurai, 1964[1]

4.1 Introduction

In this chapter, we will discuss the procedure for constructing a gauge theory of the Yang-Mills type based on a general non-Abelian gauge group. Our purpose is to derive the equations of motion for a non-Abelian gauge theory and compare them with the more familiar equations of electromagnetism. To accomplish this goal, we need to introduce some basic concepts from the Lagrangian formalism of field theory. The purpose of the Lagrangian is to couple the gauge field to the sources and particles.

4.2 Building a Gauge Model

Clearly the most desirable way to obtain the correct equations of motion for a non-Abelian gauge theory would be to use the same physical arguments as in electrodynamics. However, no one has yet directly measured the non-Abelian analogues of the famous equations of Coulomb, Ampere and Faraday. Thus we have to use a more abstract approach and write down what we believe to be the most suitable Lagrangian so that the correct equations of motion are obtained from Hamilton's principle. Let us assume for pedago-

[1] J. J. Sakurai, *Invariance Principles and Elementary Particles* (Princeton University, Princeton, New Jersey, 1964).

gical purposes that we have no *a priori* knowledge of the correct form of the equations for a Yang-Mills gauge theory. We must then begin at the same point as Yang and Mills and see how to build the Lagrangian by using symmetry principles and imposing very general physical requirements. By starting at this basic level, we can better understand how much of the theory is uniquely determined and how much has to be assumed and put in "by hand".

A general feature of the Lagrangian which is familiar from classical physics is that it must contain terms which describe the difference between the kinetic and potential energies of the system. This leads to the usual equations of motion. The new requirement is that the energy terms must be invariant under local non-Abelian gauge transformations. *A priori*, local gauge invariance is not a simple constraint to impose on a Lagrangian. One cannot simply take the absolute values of wavefunctions because a complication arises from the different behaviour of particle and gauge fields under gauge transformations. We saw in chapter III that the gauge potential transforms in a non-covariant manner,

$$A_\mu \longrightarrow UA_\mu U^{-1} - \frac{i}{q}(\partial_\mu U)U^{-1} \quad . \tag{IV - 1}$$

The extra term in the potential must be cancelled so that the Lagrangian remains unchanged. However, this cancellation should not be done by introducing *ad hoc* counter terms in the Lagrangian but it should occur in a natural way that leads to the correct equations of motion.

To understand how the above requirements are satisfied, let us examine the Lagrangian for the familiar interaction between an electron field ψ and the electromagnetic potential A_μ. The Lagrangian can be written

$$\mathcal{L} = i\bar{\psi}\gamma^\mu D_\mu\psi - \tfrac{1}{4}F^{\mu\nu}F_{\mu\nu} - m\bar{\psi}\psi \quad . \tag{IV - 2}$$

From the gauge transformation laws derived in chapter III, it is easy to verify that each term in (IV - 2) is separately gauge invariant. The first term in the Lagrangian involving the covariant derivative gives the kinetic energy of the electron. The second term is the familiar form for the energy density contained in the electromagnetic field and the last term gives the mass of the electron. Applying the Euler-Lagrange equations to (IV - 2) yields the Dirac and Maxwell equations:

$$i\gamma^\mu D_\mu \psi = m\psi \quad , \tag{IV - 3a}$$

$$\partial^\mu F_{\mu\nu} = j_\nu \quad , \tag{IV - 3b}$$

$$j_\nu = q\bar{\psi}\gamma_\nu\psi \quad . \tag{IV - 3c}$$

We see that the functional form of the Lagrangian is constrained by the need to obtain the correct equations of motion. The Dirac equation comes from the kinetic energy term and the electron mass term. The field energy term in the Lagrangian gives rise to the left-hand side of the Maxwell equation (IV - 3b) while the current j_ν comes from the kinetic energy term. The Lagrangian (IV - 2) has all the necessary terms to give the equations of motion and no more. How then does the unwanted term in the potential transformation (IV - 1) get cancelled in the Lagrangian? The answer, of course, is that the cancellation is assured by the form of the interaction in the kinetic energy term. As we saw in chapter III, the covariant derivative or minimal coupling guarantees that the extra term in (IV - 1) is exactly cancelled by the change in phase of the wavefunction as the particle moves through the field.

The example of the electromagnetic interaction provides us with a guide for generalizing the Lagrangian to a non-Abelian gauge theory. First of all, it is clear that the kinetic energy and mass terms are determined by the form of the covariant derivative and the Dirac equation. Since the covariant derivative is derived from the basic concept of the connection, it is independent of any particular gauge theory Lagrangian. Similarly, we would like to believe that the Dirac equation holds for any relativistic spin 1/2 field, including nucleons and quarks. Given these two requirements, it is reasonable to conclude that the form of the kinetic energy and mass terms remain the same for a general non-Abelian gauge theory. We cannot be quite so definite about the term in the Lagrangian for the energy density of the gauge field. This term gives rise to the left-hand side of the Maxwell equation (IV - 3b). Since we know that the field tensor $F_{\mu\nu}$ for a non-Abelian gauge symmetry has extra non-commuting terms, it is not obvious that the field equations for a Yang-Mills theory should have the same form as Maxwell's equations. Yang and Mills used the same form for the field energy as the electromagnetic case. We should nevertheless keep in mind that it may not be the only physically allowed choice.

From the preceding discussion, we can see that the electromagnetic Lagrangian provides the most reasonable starting point for a general non-Abelian theory. It is an example of using Occam's razor[a] because the electromagnetic Lagrangian is the simplest case with all the necessary properties. However, it is important to note that one is making some far reaching assumptions by modelling the Yang-Mills Lagrangian directly after that of electromagnetism. All classes of theoretical models which do not resemble electromagnetism are being excluded even if they might satisfy the general properties for a gauge theory discussed above. Whether this is reasonable can only be determined by testing the theoretical predictions against experiments.

4.3 The Problem of Mass

What is conspicuously lacking from the Lagrangian is any explicit term which provides a mass for the gauge field in the same way that the third term in (IV - 2) gives the mass of the electron. Obviously no such term is needed for the photon. However, we have earlier noted in Chapter II that the usual prescription for building a gauge field mass into the Lagrangian explicitly violates gauge invariance. Let us now see why this is always true. The form of the mass term for the potential field A_μ is determined by the require-

[a]*Entia non sunt multiplicanda praeter necessitatem* (A satisfactory proposition should contain no unnecessary complications)

William of Occam, 14th Century Oxford philosopher

ment that each component of A_μ satisfy the Klein-Gordon wave equation,

$$(\partial^\nu \partial_\nu - m^2)A_\mu = 0 \quad . \tag{IV - 4}$$

Thus the mass term has the form

$$m^2 A^\mu A_\mu \quad . \tag{IV - 5}$$

This term clearly is not gauge invariant because of the extra term that arises from the transformation of the potential in (IV - 1). We noted earlier for the kinetic energy in (IV - 2) how the extra term from the potential is exactly cancelled in the covariant derivative. Unfortunately, no such cancellation occurs for the extra terms in (IV - 5). Thus, the gauge fields are required to be massless in a Yang-Mills theory.

4.4 The Yang-Mills Equations

We can now obtain the equations of motion for a general non-Abelian gauge theory by replacing the fields in the Lagrangian (IV - 2) with new fields that carry the desired internal quantum numbers. In the original Yang-Mills theory, the field ψ is an isospin wavefunction with two components corresponding to the up and down states. For a general gauge group G, we can consider the field ψ to be an n-component vector in the internal space. The potential A_μ is now an internal operator given by a linear combination of the generators of the gauge group G. The Maxwell field tensor $F_{\mu\nu}$ is also an internal space operator and has the non-Abelian form derived in chapter III. In general, the individual components of A_μ and $F_{\mu\nu}$ will not commute among themselves since the group is non-Abelian.

It is now a straightforward exercise to obtain the equations of motion from Hamilton's principle. The only technical difficulty in applying the Euler-Lagrange equations is that one must be careful to preserve the order of the potential field components since they do not commute. Treating the potential A_μ as the independent variable gives the Yang-Mills field equation,

$$\partial^\mu F_{\mu\nu} - iq[A^\mu, F_{\mu\nu}] = j_\nu \quad , \tag{IV - 6}$$

where the current j_ν is given by

$$(j_\nu)^k = q\bar{\psi}\gamma_\nu F^k \psi \quad . \tag{IV - 7}$$

It is important to note that the current j_ν is a linear combination of the group generators F^k like the gauge fields. Thus, the current also is an internal operator and can be considered to have a direction in the internal space.

The Yang-Mills field equation (IV - 6) is the non-Abelian analogue of the inhomogeneous Maxwell equation (IV - 3b). We see that the equations are not the same despite the fact that the Lagrangians have the same functional form. There is a new commutator term in the Yang-Mills equation which has no counterpart in the electromagnetic equation. This new term occurs because the components of the gauge potential field do not commute with each other at different space-time points. This complication has to be taken into account when calculating the derivative of a gauge field like $F_{\mu\nu}$. The Euler-Lagrange equation gives the correct form of the non-Abelian divergence of $F_{\mu\nu}$, but it also tends to obscure the origin and the physical significance of the new term. In the next chapter, we will investigate further just how the usual operations of taking the divergence and gradient must be generalized for quantities in a non-Abelian gauge theory.

4.5 Particle Equation of Motion

As a first step toward understanding the physics of the Yang-Mills theory, let us examine the behaviour of a single particle in an external gauge field. The equation of motion for a particle field ψ is the Dirac equation just as in electromagnetism. However, in the Yang-Mills theory, the Dirac equation has some new features because it describes the behaviour of the particle in the internal space as well. In the following discussion, we will see how the external and internal motion can be easily understood by using some well known techniques from basic quantum mechanics.[2]

[2] S. K. Wong, *Nuovo Cimento* **65**A, 689 (1970).

In our semiclassical formalism, we can make use of Ehrenfest's theorem[3] which states that the mean value of an operator Q obeys the following law of motion:

$$i\frac{d}{dt}\langle Q\rangle = \langle [Q, H]\rangle \quad , \qquad \text{(IV - 8)}$$

where H is the Hamiltonian. Since we started from the Dirac equation, the appropriate Hamiltonian can be written in the usual form[3],

$$H = \boldsymbol{\alpha}\cdot(\mathbf{p} - q\mathbf{A}) + m\beta - iqA_0 \quad , \qquad \text{(IV - 9)}$$

where $\boldsymbol{\alpha}$ and β are the Dirac matrices. Let us first calculate the usual kinematical quantities $d\mathbf{x}/dt$ and $d\mathbf{P}/dt$, where $\mathbf{P} = \mathbf{p} - q\mathbf{A}$ is the canonical momentum. Since $\mathbf{x}$ commutes with all of the group generators, we obtain the familiar result

$$\frac{d}{dt}\mathbf{x} = \boldsymbol{\alpha} \quad . \qquad \text{(IV - 10)}$$

The calculation for the canonical momentum is slightly more complicated because we must use the commutation relation of the group generators for the components of the gauge field A_μ. In addition, the operator $\mathbf{p}$ in the Hamiltonian also acts as a derivative on A_μ. After some algebra, we obtain

$$\frac{d}{dt}P_k = q(\alpha_j \partial_k A_j + i\partial_k A_0) \quad . \qquad \text{(IV - 11)}$$

If we express the field tensor $F_{\mu\nu}$ in terms of the electric and magnetic fields,

$$F_{0i} = E_i \quad , \qquad \text{(IV - 12a)}$$

$$F_{ij} = \epsilon_{ijk}B_k \quad , \qquad \text{(IV - 12b)}$$

it is easy to see that (IV - 11) is just the non-Abelian version of the familiar Lorentz force law.

In order to study the particle motion in the internal space, we

[3] A. Messiah, *Quantum Mechanics* (J. Wiley, New York, 1961).

note that the internal quantum numbers are eigenvalues of the group generators. Thus we can apply (IV - 8) directly to a group generator F_k and obtain

$$\frac{\mathrm{d}}{\mathrm{d}t}\langle F_k\rangle = \left\{ q\boldsymbol{\alpha}\cdot\mathbf{A}\times F_k + \mathrm{i}qA_0\times F_k \right\}$$

$$= q\left\{ \frac{\mathrm{d}\mathbf{x}}{\mathrm{d}t}\cdot\mathbf{A} + \mathrm{i}A_0 \right\}\times F_k \quad , \qquad \text{(IV - 13)}$$

where the cross product indicates a commutation relation of F_k and the group generators implicit in the potential field components. To understand the meaning of this equation, let us compare it with the familiar expression for the Thomas precession of spin in a magnetic field. In the rest frame, the spin polarization for a point electron obeys the equation

$$\frac{\mathrm{d}}{\mathrm{d}t}\mathbf{S} = \frac{e}{mc}\mathbf{S}\times\mathbf{B} \quad . \qquad \text{(IV - 14)}$$

The corresponding rest frame expression for $\langle F_k\rangle$ from (IV - 13) is

$$\frac{\mathrm{d}}{\mathrm{d}t}\langle F_k\rangle = \mathrm{i}qA_0\times F_k \quad . \qquad \text{(IV - 15)}$$

Comparing these two equations, we see that the mean value of the group generator F_k undergoes a precessional motion as the particle moves through the gauge field. We stated earlier in chapter III that the gauge potential rotates the internal space direction of the particle; we now see directly from (IV - 15) just how this occurs. We can also see from the above equations that the internal space rotation differs from ordinary spin precession in an essential way. If the magnetic field is zero in (IV - 14), there is no spin precession. However, there will still be an internal space rotation as along as A_0 does not vanish. This situation corresponds exactly to the Aharonov-Bohm effect discussed in chapter II. The internal space rotation describes the change in phase of the electron beams as they moved around the solenoid.

The preceding discussion also leads very simply to the conservation laws for the internal space variables. From equation (IV - 8), it is evident that the only internal quantities which will be conserved are those which commute with the Hamiltonian. Since the gauge fields in the Hamiltonian are linear combinations of the group generators, the conserved quantities therefore must commute with all of the generators. For the SU(2) group in the Yang-Mills theory, we know that the square of the total isospin commutes with the three isospin generators and is conserved. For larger gauge symmetry groups, such as SU(3), there will be more than one quantity which commutes with the generators and is conserved. In any case, we can see that the conservation laws are independent of the internal space motion and depend only on the symmetry group itself.

CHAPTER V

THE MAXWELL EQUATIONS

I also found that several of the most fertile methods of research discovered by the mathematicians could be expressed much better in terms of the ideas derived from Faraday than in their original form.

James Clerk Maxwell, 1873[1]

5.1 Introduction

One of the most important differences between electromagnetism and Yang-Mills theory is that the role of the photon is played by a set of local non-Abelian potential fields with entirely new properties. Although these potentials are massless vector fields like the photon, they are also internal-space operators which do not commute with each other. In this chapter, we will study the salient features of the gauge fields through the Yang-Mills equation (IV - 6) and compare them with the Maxwell equations. We shall see why the commutation properties of the gauge fields are the reason for most of the interesting differences between electromagnetism and non-Abelian gauge theory.

5.2 The Non-Abelian Divergence

In chapter IV, we noted that the Yang-Mills field equation

$$D^{\mu} F_{\mu\nu} = \partial^{\mu} F_{\mu\nu} - iq[A^{\mu}, F_{\mu\nu}] = j_{\nu} \qquad \text{(V - 1)}$$

has a new term, involving a commutator of gauge fields, which has no counterpart in the inhomogeneous Maxwell equation. This term arises in a non-Abelian theory because the gauge field com-

[1]J. C. Maxwell, *A Treatise on Electricity and Magnetism* (Dover, New York, 1954).

ponents do not commute with each other at different points in space-time. In order to calculate the divergence of the field $F_{\mu\nu}$ in the Yang-Mills equation, we must know its value at two nearby positions x and $x + dx$ so that we can find the rate of change. However, a gauge field like $F_{\mu\nu}$ or any of its derivatives is not only a function of space-time position but also has a direction in the internal space which can change between x and $x + dx$. Since the gauge fields do not commute, the internal space direction of $F_{\mu\nu}$ at $x + dx$ will depend on the particular path taken between x and $x + dx$. This is a complication which does not occur when taking ordinary partial derivatives in electromagnetism. Thus, in order to understand the origin of the new term in the Yang-Mills equation (V - 1), we must first study the more general problem of how to compute the divergence or gradient in a non-Abelian gauge theory. We will consider only the specific case of the Maxwell tensor $F_{\mu\nu}$ but the following discussion is applicable to other cases.

To see how $F_{\mu\nu}$ changes between x and $x + dx$, we can make use of a familiar pedagogical device from electromagnetism. The basic idea is to compute the flux through a small surface by using a test charge which is transported around the boundary of the surface. As the test charge moves along the boundary as shown in Fig. (5.1), the internal space direction will rotate and the phase of the test charge wavefunction will change. The net change of the phase can be related by Stokes' theorem to the total flux through the surface and to the value of the Maxwell tensor.

The derivative of the flux can be computed by performing the

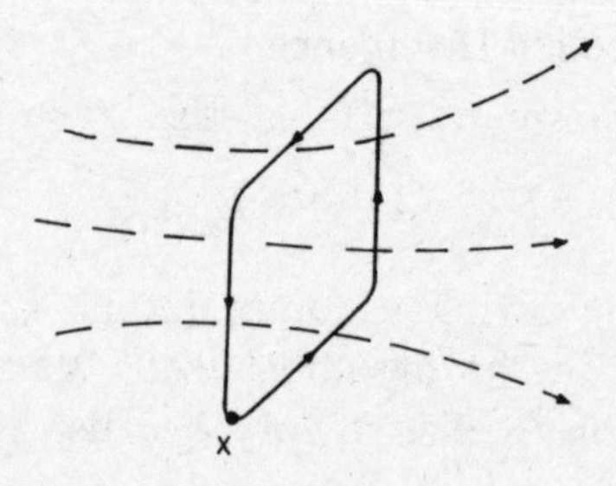

Fig. 5.1 Test charge is moved around the boundary of a surface to measure the flux though the surface.

above flux measurement at the two locations x and $x + \mathrm{d}x$. After measuring the flux at x, the test charge must be moved from x to $x + \mathrm{d}x$. The flux is then measured at $x + \mathrm{d}x$ and the test charge is moved back again along the same path to x. The net change in phase of the test charge then allows us to calculate the derivative of the flux at x. The complete sequence of movements of the test charge is shown in Fig. (5.2). This procedure ensures that the total change of the Maxwell tensor $F_{\mu\nu}$, both externally in space-time and in the internal space, are included in the derivative in a self-consistent way.

For each step in Fig. (5.2), the change in phase of the test charge can be described by the effect of an infinitesimal gauge transformation acting on its wavefunction. For the spatial displacement $x \rightarrow x + \mathrm{d}x$, we see from the discussion in chapter III that the transformation is given to first order in $\mathrm{d}x$ by[a]

$$U_x(\mathrm{d}x) = 1 - \mathrm{i}qA\cdot\mathrm{d}x \quad . \qquad \text{(V - 2)}$$

We first use this formula to evaluate the phase change around the boundary of the small flux surface. By summing up the successive contributions from all sides of the boundary, we see that the net change in the wavefunction of the test charge after it has returned to x is

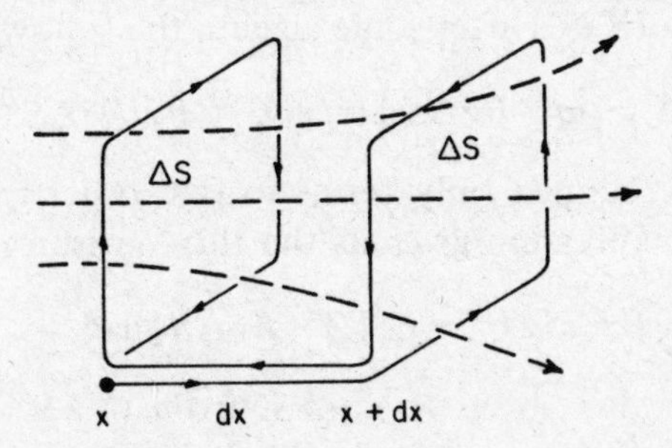

Fig. 5.2 Complete path taken by the test charge for the measurement of the non-Abelian divergence of the Maxwell field tensor $F_{\mu\nu}$ at the position x.

[a]To avoid the excessive use of indices, we adopt the simplified notation $\mathrm{A}\cdot\mathrm{dx} = \mathrm{A}_\mu \mathrm{dx}^\mu$ and $\mathrm{F}\cdot\mathrm{dS} = \mathrm{F}_{\mu\nu}\mathrm{dS}^{\mu\nu}$ in equations (V - 2) through (V - 10).

$$d\psi = -iq \sum A \cdot dx\, \psi \quad . \qquad (V-3)$$

The summation over the vector potential can be related to the Maxwell tensor $F_{\mu\nu}$ at x by using Stokes' theorem,

$$\sum A \cdot dx = \iint F \cdot dS \simeq \widetilde{F} \cdot \Delta S \quad , \qquad (V-4)$$

where $\widetilde{F}$ is the average value over the infinitesimal surface ΔS. Using Eq. (V - 4) in (V - 3) then gives the effective gauge transformation for one circuit around the boundary

$$U_x = 1 - iq\widetilde{F}(x) \cdot \Delta S \quad . \qquad (V-5)$$

The infinitesimal gauge transformations corresponding to the complete sequence of movements in Fig. (5.2) can be written as follows. Moving the test charge from x to $x + dx$, transporting it once around the boundary of the surface at $x + dx$, and then bringing the test charge back again to x corresponds to the product of three gauge transformations

$$[1 - iqA \cdot dx]\,[1 - iq\widetilde{F}(x + dx) \cdot \Delta S]\,[1 + iqA \cdot dx] \quad . \qquad (V-6)$$

We expand $\widetilde{F}(x + dx) \cdot \Delta S$ to the first order in dx,

$$\widetilde{F}(x + dx) \cdot \Delta S = \widetilde{F}(x) \cdot \Delta S + \partial_\mu \widetilde{F} \cdot \Delta S\, dx^\mu \quad , \qquad (V-7)$$

and obtain from (V - 6) after some algebra the following,

$$1 - iq\widetilde{F} \cdot \Delta S - iq\partial_\mu \widetilde{F} \cdot \Delta S\, dx^\mu - q^2 [A_\mu, \widetilde{F} \cdot \Delta S]\, dx^\mu \quad , \qquad (V-8)$$

where we have retained only terms to first order in dx. The phase change in this expression gives us the flux measurement at $x + dx$,

$$\widetilde{F} \cdot \Delta S + \partial_\mu \widetilde{F} \cdot \Delta S\, dx^\mu - iq[A_\mu, \widetilde{F} \cdot \Delta S]\, dx^\mu \quad . \qquad (V-9)$$

Subtracting the flux at x, $\widetilde{F}(x) \cdot \Delta S$, from (V - 9) then yields the net change of the flux between x and $x + dx$,

$$\left\{ \partial_\mu \widetilde{F} \cdot \Delta S - iq[A_\mu, \widetilde{F} \cdot \Delta S] \right\} dx^\mu \quad . \qquad (V-10)$$

We can easily see by inspection that the derivative of the Maxwell field tensor is

$$D^{\lambda}F_{\mu\nu} = \partial^{\lambda}F_{\mu\nu} - iq[A^{\lambda}, F_{\mu\nu}] \quad . \tag{V - 11}$$

Contracting the λ and μ indices in Eq. (V - 11) finally gives us the non-Abelian version of the divergence of F which we can compare with the Yang-Mills equation (V - 1).

We now see that the new commutator term in the Yang-Mills equation can be considered as a part of the divergence itself and is a direct consequnce of the fact that $F_{\mu\nu}$ and A^{λ} do not commute. Our derivation of the term using purely geometrical arguments in the internal space shows that it does not depend on the details of the Lagrangian which we wrote down in chapter IV or on the use of the Euler-Lagrange equations. Of course, the essential purpose of the Lagrangian is to equate the divergence of $F_{\mu\nu}$ to the source current in both electromagnetism and non-Abelian gauge theory. We see that we might have obtained the Yang-Mills equation from the inhomogeneous Maxwell equation by simply replacing the ordinary partial derivative with the gauge-covariant derivative. However we did not assume *a priori* that we knew the form of the non-Abelian divergence of $F_{\mu\nu}$, which is different from the gauge-covariant derivative of a wavefunction. We also gained some physical insight into the Yang-Mills equation by relating it to the familiar idea of flux measurement.

5.3 The Second Maxwell Equation and Charge Conservation

Let us now examine the non-Abelian versions of the homogeneous Maxwell equation and the law of charge conservation. In ordinary electrodynamics, the homogeneous Maxwell equation is written

$$\partial_{\mu}F_{\nu\lambda} + \partial_{\lambda}F_{\mu\nu} + \partial_{\nu}F_{\lambda\mu} = 0 \quad . \tag{V - 12}$$

This equation is not derived from a Lagrangian and only depends on the space-time properties of the Maxwell tensor. Thus, we can use the results of the previous section and replace the partial derivatives with the gauge covariant derivative (V - 11) so that the corresponding non-Abelian equation becomes

$$D_\mu F_{\nu\lambda} + D_\lambda F_{\mu\nu} + D_\nu F_{\lambda\mu} = 0 \quad . \tag{V - 13}$$

In order to better understand this equation, we can interpret its meaning in the internal space by again using a moving test charge. We associate each term in Eq. (V - 13) with one of the paths shown in Fig. (5.3). As in the preceding section, each path measures the flux difference between opposite faces of an infinitesimal test volume. When the contributions from the three paths are summed according to Eq. (V - 13), the net change in the internal phase must vanish. It is easy to see from Fig. (5.3) that the phase changes along the paths all cancel. For example, consider the path segment AB in Fig. (5.3a) and the segment CD in Fig. (5.3c). The phase changes along these two segments are equal in magnitude but opposite in sign. The same type of cancellation occurs for all other segments.

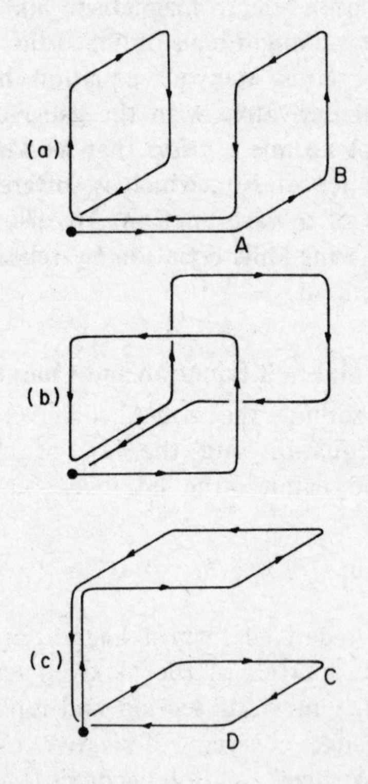

Fig. 5.3 Set of three paths for the test charge which correspond to the three terms in the homogeneous non-Abelian Maxwell equation.

Thus the net phase change from the three paths is zero as required by Eq. (V - 13). Since the phase change provides a measure of the flux difference, we see that the physical interpretation of Eq. (V - 13) is the same as that in QED, namely, that the total flux out of the test volume vanishes.

The conservation of electric charge in QED is commonly expressed by

$$\partial^\nu(\partial^\mu F_{\mu\nu}) = \partial^\nu j_\nu = 0 \quad . \tag{V - 14}$$

We can generalize this equation to the conservation of non-Abelian charge by again replacing the ordinary divergences to obtain

$$D^\nu(D^\mu F_{\mu\nu}) = D^\nu j_\nu = 0 \quad . \tag{V - 15}$$

This new conservation law does not simply describe the conservation of charge carried by the source current as in QED. Since the gauge fields have internal degrees of freedom, (V - 15) must also apply to the gauge fields alone when there is no explicit source current. In QED, (V - 14) is trivially satisfied when there are no sources because the field can be only a constant. This is not true in non-Abelian gauge theory because the source current j_ν is not locally conserved like the electromagnetic current. By local conservation, we mean that the ordinary spatial divergence of j_ν also vanishes as in QED. Since this is not true, we cannot simply apply Gauss's law in its usual integral form in non-Abelian theory to calculate the net charge contained within a volume in the same way that we normally do in electrodynamics. In order to define a conserved non-Abelian current, we can re-write the Yang-Mills equation (V - 1) as follows,

$$\partial^\mu F_{\mu\nu} = J_\nu \quad , \tag{V - 16}$$

where J_ν is defined to be the new total current density

$$J_\nu = j_\nu + iq[A^\mu, F_{\mu\nu}] \quad . \tag{V - 17}$$

It is easy to see that J_ν is a conserved current because J_ν is an ordinary divergence of an antisymmetric tensor just as in QED. The total non-Abelian charge can be defined by the usual volume integral,

$$Q = \int d^3x J_0 \quad . \tag{V - 18}$$

This total charge is a time-independent scalar quantity but it is also a vector in the internal space. Thus a gauge transformation can rotate the non-Abelian charge unlike the case of electric charge.

The total current J_ν appears to have the useful property of explicitly separating the charge contributions from the source and the field. However, this separation is not gauge invariant and therefore cannot be physical. Under a local gauge transformation, the covariant divergence of $F_{\mu\nu}$ transforms like $F_{\mu\nu}$ itself,

$$F_{\mu\nu} \longrightarrow UF_{\mu\nu}U^{-1} \quad , \tag{V - 19a}$$

$$D^\mu F_{\mu\nu} \longrightarrow UD^\mu F_{\mu\nu}U^{-1} \quad . \tag{V - 19b}$$

However, the ordinary divergence of $F_{\mu\nu}$ transforms as

$$\partial^\mu F_{\mu\nu} \longrightarrow U\partial^\mu F_{\mu\nu}U^{-1} + U[U^{-1}\partial^\mu U, F_{\mu\nu}]U^{-1} \quad . \tag{V - 20}$$

The extra term in (V - 20) arises from the well-known, non-gauge covariant transformation property of the gauge field itself,

$$A_\mu \longrightarrow UA_\mu U^{-1} - \frac{i}{q}(\partial_\mu U)U^{-1} \quad . \tag{V - 21}$$

Therefore unlike the source current, the current carried by the field is not gauge invariant due to the explicit presence of the vector potential A_μ in the commutator term. In fact it is easy to show that the commutator transforms in just the right way under local gauge transformations so as to cancel the extra term which arises in Eq. (V - 20), thus ensuring that the source current j_ν is gauge invariant. This non-invariant property of the field current means that the charge density of the field is not locally defined. It can take on any value at a given point in space-time because the exact gauge symmetry will always allow us to perform arbitrary gauge transformation on the potential A_μ. The charge density of the field can even be set locally to zero without affecting the physics.

This is the same as choosing a particular gauge. Thus if we try to calculate the charge of the field itself within some source-free region using Gauss's law, it will appear that the result of the calculation is not conserved or even well-defined. This does not mean that the gauge field does not carry charge or that charge is not conserved. It does mean that only the total charge is well defined in a non-Abelian gauge theory and that Gauss's integral equation must always be applied to both the source and field to obtain a conserved charge.

5.4 Maxwell's Equations and Superposition

Let us now rewrite the non-Abelian equations in terms of the electric and magnetic fields in order to compare them directly with the individual QED Maxwell equations. In analogy with electrodynamics, the electric and magnetic fields are defined as follows:

$$F_{0i} = E_i \ , \qquad F_{ij} = \epsilon_{ijk} B_k \quad . \qquad \text{(V - 22)}$$

The time component of the Yang-Mills equation (V - 1) gives us the non-Abelian version of Gauss' law

$$\nabla \cdot \mathbf{E} + iq(\mathbf{A} \cdot \mathbf{E} - \mathbf{E} \cdot \mathbf{A}) = j_0 \quad . \qquad \text{(V - 23)}$$

The space components give us the non-Abelian Ampere's law

$$\frac{\partial \mathbf{E}}{\partial t} - \nabla \times \mathbf{B} + iq[A_0, \mathbf{E}] - iq(\mathbf{A} \times \mathbf{B} - \mathbf{B} \times \mathbf{A}) = \mathbf{j} \quad . \qquad \text{(V - 24)}$$

To extract the non-Abelian version of Faraday's law from Eq. (V - 13), one chooses two space indices and one time index and obtains

$$\nabla \times \mathbf{E} + \frac{\partial \mathbf{B}}{\partial t} + iq[A_0, \mathbf{B}] + iq(\mathbf{A} \times \mathbf{E} - \mathbf{E} \times \mathbf{A}) = 0 \ . \qquad \text{(V - 25)}$$

Choosing all three space indices in Eq. (V - 13) gives the non-Abelian version of the "no monopole" equation for the magnetic field

$$\nabla \cdot \mathbf{B} + iq(\mathbf{A} \cdot \mathbf{B} - \mathbf{B} \cdot \mathbf{A}) = 0 \quad . \qquad \text{(V - 26)}$$

The current conservation law (V - 15) can be written as

$$\nabla \cdot \mathbf{j} - \frac{\partial j_0}{\partial t} - iq(\mathbf{A} \cdot \mathbf{j} - \mathbf{j} \cdot \mathbf{A}) - iq[A_0, j_0] = 0 \quad . \qquad \text{(V - 27)}$$

We see that the non-Abelian versions of Maxwell's equations contain contributions which are identical in form to the ordinary electrodynamic Maxwell equations. It is clear that these equations will reduce to the QED equations if all the fields commute. However, a more important reason for writing out the non-Abelian Maxwell equations is that they point out explicitly the most serious differences between QED and Yang-Mills gauge theory. The extra commutator terms show that the electric and magnetic fields in a general non-Abelian gauge theory do not obey the principle of superposition. For example, let us consider two sources which produce electric and magnetic fields that individually satisfy Maxwell's equations. The usual vector sum of the fields will not in general be a solution to these equations. The equations will contain mixed commutator terms like $(\mathbf{A}\cdot\mathbf{E} - \mathbf{E}\cdot\mathbf{A})$ which are non-zero because the field **A** of one charge does not commute with the field **E** of the other charge. In addition, a general solution for the two sources may not factorize into a sum of two solutions, one for each of the separate fields.

Superposition is such a fundamental concept in electrodynamics that the violation of the principle in non-Abelian gauge theory raises some serious questions. Many of the most useful techniques in electrodynamics for solving boundary value or wave problems are based on some form of linear expansion or superposition of solutions. Clearly such methods cannot be carried over directly to Yang-Mills theory. In the following sections, we will briefly discuss two of the most familiar problems in electrodynamics, namely the Coulomb force between stationary charges and the simple plane wave, and see how they are affected by the lack of superposition.

5.5 Charges in Yang-Mills Theory

One of the simplest problems in electrodynamics is to find the electric field produced by a system of point charges at rest. This problem is of interest in non-Abelian gauge theory because it is the starting point for understanding the nature of the forces acting between elementary particles. We cannot, of course, expect a semiclassical formalism to describe the behaviour of real particles but it can show us the type of problems caused by the noncommutativity of the gauge field.

The electric field of an isolated charge at rest was investigated in Yang-Mills theory by Ikeda and Miyachi[2] and Loos[3]. The original motivation was to see if a solution could be found for which the quanta of the Yang-Mills field might have a non-zero mass produced by the non-linear self-interactions of the gauge fields. However, the most general electric field of a single charge in Yang-Mills theory was found to be a Coulomb field just as in electrodynamics.

In order to solve for the gauge field, we first require that the solution be spherically symmetric about the charge. It is straightforward to see that the field can be written in the general form

$$\mathbf{A} = \frac{\mathbf{x}}{r} f(r, t) \quad , \qquad A_0 = \mathrm{i} g(r, t) \quad , \qquad \text{(V - 28)}$$

where $f(r, t)$ and $g(r, t)$ are arbitrary functions of the radius r and time t. For a purely static solution, we can also choose $g(r, t)$ to be zero and solve only for $f(r)$. The Yang-Mills equation in the source-free region reduces to the simple form

$$\frac{\partial f}{\partial r} + \frac{2f}{r} = 0 \quad , \qquad \text{(V - 29)}$$

which can be easily integrated to yield the Coulomb solution

$$f(r) = \frac{\lambda}{r^2} \quad , \qquad \text{(V - 30)}$$

where the integration constant λ is the non-Abelian charge vector of the field. The general case with $g(r, t)$ non-zero can be solved analytically[2] but our purpose is only to see that the field of a single Yang-Mills charge does indeed have the Coulomb form.

It is clear that the Coulomb solution is a consequence of the imposed requirement of spherical symmetry since nothing else was

[2] M. Ikeda and Y. Miyachi, *Prog. Theor. Phys.* 27, 474 (1962).
[3] H. G. Loos, *Nucl. Phys.* 72, 677 (1965).

said about the properties of the source charge. This symmetry reduces the non-Abelian Maxwell equations down to the same form as electrodynamics. For a single isolated point charge, the spherical symmetry permits us to choose the charge vector of the field completely arbitrarily at every point in space-time. Thus the choice of charge direction cannot have any physical effect, and the internal space dependence is effectively decoupled from the space-time dependence of the field. For convenience, we can therefore adopt the convention that the charge vector of the field has the same fixed direction at every point. This is just a particular choice of gauge which makes it easy to see that $\mathbf{A}$, its derivatives and the fields constructed from them all point in the same internal space direction and therefore commute with each other.

The single Yang-Mills charge is an example of one case where it is particularly convenient to be able to choose a gauge where the current density of the gauge field is zero. For a Coulomb field, the spherical symmetry means that a gauge transformation can be found which will effectively transfer all of the charge carried by the field onto the source. The total charge therefore can be determined from the source density alone by a volume integral just as in electromagnetism.

Unfortunately, the simple arguments used for the solution of a single charge cannot be extended to several charges. When there are two or more charges, each at a different position, we cannot deduce from symmetry arguments alone that all of the troublesome commutators vanish. Attempts to find general analytic solutions for bound states involving an arbitrary configuration of charges in Yang-Mills theory have so far been unsuccessful. Even the non-Abelian version of the positronium atom has not been solved completely. To illustrate the difficulties arising from the non-linear nature of the Yang-Mills equations, let us consider the simple example of two point charges at rest[4,5]. We can express

[4] P. Sikivie and N. Weiss, *Phys. Rev.* **D18**, 3809 (1978).
[5] I. B. Khriplovich, *Sov. Phys. JETP* **47**, 18 (1978).

the charges as follows:

$$q_1 = qF_1\,\delta(\mathbf{x}-\mathbf{x}_1)\quad ,$$

$$q_2 = qF_2\,\delta(\mathbf{x}-\mathbf{x}_2)\quad , \tag{V - 31}$$

where F_1 and F_2 are generators of the gauge symmetry group. The generators indicate the directions of the charges in the internal symmetry space. To see the effects of the non-linearity, let us deliberately write the potential field produced by these two point charges as the linear combination

$$A_\mu = A_\mu^1 F_1 + A_\mu^2 F_2 \quad . \tag{V - 32}$$

We saw previously that the potential will rotate the internal direction of a charge which is moving through the field. We can now use the Yang-Mills equation to see how the potential (V - 32) affects the internal direction of a charge which is at rest. Since the generators indicate the direction, we need to apply the Yang-Mills equation to the F_j themselves,

$$\begin{aligned} D_\mu F_j &= \partial_\mu F_j + q[A_\mu, F_j] \\ &= \partial_\mu F_j + qA_\mu^k [F_k, F_j] \qquad j, k = 1, 2 \quad , \end{aligned} \tag{V - 33}$$

which represents a set of two coupled equations for the two charges. The source terms vanish in these equations because the field from the point charge at x_j acts on the internal direction of the charge at x_j. The change with respect to time of F_j is thus given by

$$\frac{\partial}{\partial t} F_j = -qA_0^k [F_k, F_j] \qquad j, k = 1, 2 \quad . \tag{V - 34}$$

This result indicates that the internal directions of the charges do not remain fixed. Even though the charges are at rest in space-time, their internal directions appear to rotate or precess as a function of time. This peculiar situation is due to the fact that we tried to fix the internal directions of the two charges with F_1 and F_2 and also assumed that the potential could be written as the linear combination

(V - 32). The result (V - 34) clearly shows us that the fields from two non-Abelian charges cannot be superposed in general.

We note from (V - 34) that physically reasonable solutions do exist for two charges when the commutator term vanishes. The equations (V - 33) become uncoupled and the gauge potential field for the two charges is simply a linear combination of Coulomb fields. The number of such solutions depends on the properties of the gauge symmetry group. For the SU(2) isotopic-spin group, the internal direction of all charges would have to be "parallel" or "anti-parallel" at all points in space-time. For larger gauge groups, there may be several different generators which commute with each others. For example in SU(3), there are two generators, F_3 and F_8, which commute. Using these two generators, one can define three different linearly independent charges.

5.6 Yang-Mills Wave Equation

The equation for a non-Abelian wave can be obtained directly from the Yang-Mills equation by setting the source current to zero. In order to compare this equation with the familiar electrodynamic case, we rewrite the equation explicitly in terms of the gauge potential

$$D^\mu F_{\mu\nu} = \partial^\mu \partial_\mu A_\nu - \partial_\nu(\partial^\mu A_\mu) + q\left\{[A^\mu, F_{\mu\nu}] + \partial^\mu [A_\mu, A_\nu]\right\} \quad . \qquad \text{(V - 35)}$$

We see that the first and second terms are identical in form with the electromagnetic wave equation. The second term can be set to zero as usual by choosing the Lorentz gauge. However, the commutator terms clearly indicate that non-Abelian waves also do not in general obey the principle of superposition.

Are there plane-wave solutions to the Yang-Mills wave equation that might be associated with free non-Abelian photons? The question of plane waves has been answered affirmatively by Coleman[6]

[6] S. Coleman, *Phys. Lett* 70B, 59 (1977a).

who has found the most general plane-wave solutions for an arbitrary non-Abelian gauge symmetry. The solutions have the form

$$A_x = A_y = 0 \quad ,$$

$$A_0 = -A_z = xf(z-t) + yg(z-t) \quad , \qquad \text{(V-36)}$$

where the velocity of light has been set equal to unity. These solutions describe plane waves moving in the positive z-direction, where f and g are arbitrary functions of $z - t$. The components of the Maxwell field tensor for these waves are

$$F_{0x} = F_{xz} = -f \quad ,$$

$$F_{0y} = F_{yz} = -g \quad ,$$

$$F_{0z} = F_{xy} = 0 \quad . \qquad \text{(V-37)}$$

It can easily be seen that these solutions satisfy the Yang-Mills wave equation. However, Coleman's plane waves also do not obey the principle of superposition for the same reasons given above.

Special cases of Coleman's plane-wave solutions can be found[7] which do satisfy superposition. The basic idea is to either choose a gauge which fixes the direction of the internal vector to be the same at all points in space or to construct gauge fields using only the special subset of group generators which commute. These special plane-wave solutions presumably could be interpreted as the classical analogues of free non-Abelian waves. However, such waves would have no self-interactions and thus might not be realistic.

Because the Yang-Mills gauge fields carry internal quantum numbers and thus can be charged unlike the photon, it has been suggested that the wave equation might have novel solutions which have no analogues in electrodynamics. One potentially interesting but highly speculative solution would be a hypothetical bound

[7]F. Melia and S. Lo, *Phys. Lett.* 77B, 71 (1978).

state of pure gauge fields ("glueball")[8]. This bound state presumably would hold itself together by some attractive force due to the internal quantum numbers carried by the gauge fields themselves. This idea is appealing because the experimental observation of such a bound state would clearly provide a unique signature of the non-Abelian properties of the gauge theory. However, Coleman[9,10] has argued that gauge field bound states are not likely to exist, at least in semiclassical gauge theory. Although the argument is not definitive, it is educational because it points out some of the limitations of a semiclassical argument. Roughly speaking, for a classical bound state to exist, every part of the gauge field should mutually attract every other nearby part. However, one can argue, in analogy with electric charges, that this would require the internal directions to be "antiparallel". This is not possible for a gauge field because continuity requires the internal directions for nearby portions of the field to point in the same direction. Thus one might say that the field is "self-repulsive" at small distances. This argument appears to rule out stable bound states of gauge fields in the semiclassical theory, but Coleman points out that it does not necessary forbid the existence of such objects in a quantum field theory. The moral of this story is that a non-Abelian gauge theory may contain entirely new types of objects never considered possible in physics before.

[8] P. Roy, *Nature* **280**, 106 (1979)
[9] S. Coleman, *Comm. Math. Phys.* **55**, 113 (1977b)
[10] S. Coleman and L. Smarr, *Comm. Math. Phys.* **56**, 1 (1977c).

CHAPTER VI

THE DIFFICULT BIRTH OF MODERN GAUGE THEORY

It is easy to imagine that in a few years the concepts of field theory will drop totally out of the vocabulary of day-to-day work in high energy physics.

F. Dyson, 1965[1]

6.1 Introduction

In retrospect, it is clear that both the original theory of Weyl and the Yang-Mills theory failed because they were too early for a gauge theory at their respective times. When Yang and Mills proposed their revolutionary idea for an SU(2) gauge theory of the strong interaction, an adequate understanding of the essential properties of the nuclear forces was still lacking. The final successful rediscovery of gauge theory had to wait nearly ten more years for several crucial experimental and theoretical developments to provide the necessary clues.

6.2 Leptons and *W* Bosons

The first important development toward the rebirth of modern gauge theory did not involve the strong interaction. Instead, it was the weak nuclear force. A major breakthrough occurred in the understanding of the weak interaction at approximately the same time as the Yang-Mills theory. After more than twenty years of difficult and often beautiful experiments, the violation of parity by the weak interaction was discovered[2,3] and the nature of the weak nuclear force was finally elucidated. A consistent theory emerged,

[1] *op.cit.* p. 34

[2] T. D. Lee and C. N. Yang, *Phys. Rev.* **104**, 254 (1956).

[3] C. S. Wu, E. Ambler, R. Hayward, D. Hoppes and R. Hudson, *Phys. Rev.* **105**, 1413 (1957).

known as the V–A theory, which successfully described all of the known weak decay processes.[4,5]

The understanding of the weak decays was complicated by the fact that most of the available experimental data came from reactions such as the beta decay of the neutron

$$n \longrightarrow p + e^- + \bar{\nu}_e \quad .$$

These weak decays involve both strongly interacting particles and "leptons" such as the electron or muon and their associated neutrinos. The first successful theory of beta decay was developed in 1934 by Enrico Fermi[6], who modelled his original theory after electrodynamics. Fermi considered beta decay to be analogous to the emission of a photon from an electromagnetic transition in an excited atom. However, as shown in Fig. (6.1), the role of the emitted photon was not played by the quantum of a new weakly interacting field but instead by the electron-antineutrino pair. The original Fermi theory eventually evolved into a more abstract theory based on the idea of interacting "currents". The new currents were generalizations of the familiar charged current encountered in the Dirac theory,

$$J_\mu = e\bar{\psi}\gamma_\mu\psi \quad , \tag{VI - 1}$$

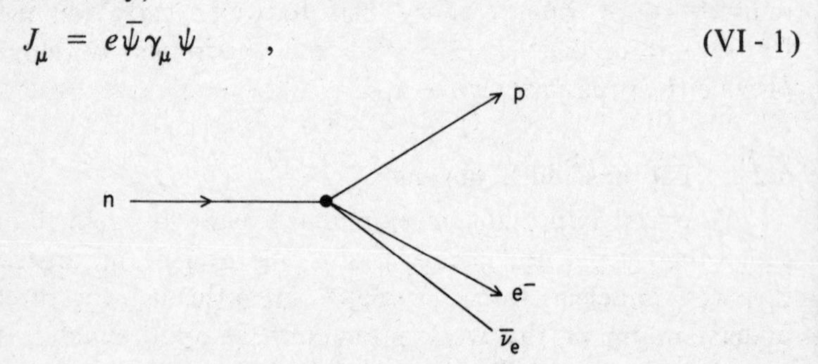

Fig. 6.1 Weak beta decay of the neutron $n \to p + e^- + \nu_e$ is represented in the Fermi theory by interaction of all particles at a single point in space-time. The emission of the $e^- - \nu_e$ pair is considered analogous to photon emission in electromagnetic transitions.

[4] R. P. Feynman and M. Gell-Mann, *Phys. Rev.* **109**, 193 (1958).
[5] R. E. Marshak and E. Sundarshan, *Phys. Rev.* **109**, 1860 (1958).
[6] E. Fermi, *Zeit Physik* **88**, 161 (1934) (An English translation is given by F. Wilson, *Am. Jour. Phys.* **36**, 1150 (1968).

where ψ is the 4-component electron spinor wavefunction and γ_μ are the 4×4 Dirac matrices. In the example of beta decay, the neutron and proton define a current J_μ which carries isotopic spin and a baryon number. The electron and its neutrino define another current J'_μ which has a well defined lepton number. These currents replaced the fields which had originally been used by Fermi to represent the particles. The weak interaction is then described by coupling the currents together through a Hamiltonian of the form

$$H = \frac{G}{\sqrt{2}} J_\mu J'_\mu \quad , \qquad \text{(VI - 2)}$$

where G is the Fermi weak coupling constant, which has a value[a] of approximately 10^{-5}. The value of G can be considered as a kind of "weak charge" analogous to the electric charge e. Since the two currents J_μ and J'_μ in (VI - 2) are coupled together at a single space-time point, the interaction was called a "local four-Fermi" coupling or a "current-current" interaction. The term "local" here means only at the same space-time point. In order to incorporate parity violation into the theory, the weak currents J_μ and J'_μ contain terms which behave like vectors (i.e. like the Dirac current) and axial vectors. These terms have approximately equal strengths but opposite signs, hence the name V–A. The V–A theory has been enormously successful in explaining essentially all of the salient features of the weak decays. In nearly all textbooks on particle physics, the V–A theory is still regarded as the definitive theory of the weak interaction.

It would appear that neither the Fermi theory nor the V–A theory bear any resemblance to a Yang-Mills gauge theory. In fact, since Fermi published his theory before the Yukawa meson theory, he did not even consider the possibility that the weak interaction might be mediated by the exchange of a new heavy photon-like quantum. The earliest proposal for a new weak quantum is credited to Yukawa himself who suggested that both the strong and weak interaction could be explained by meson exchange. The π-meson, which was predicted by Yukawa, did not turn out to be

[a]The value of G is 1.4×10^{-49} erg cm^3 or 10^{-5} $\hbar c\,(\hbar/Mc)^2$ where M is the proton mass.

the desired weak quantum. However, Yukawa's meson theory motivated others[7,8] to develop the idea of meson exchange in greater detail. In close analogy with electrodynamics it was hypothesized that the weak decay was mediated by the exchange of a new spin-1 quantum, shown in Fig. (6.2), which was called the "intermediate vector boson" or later the "*W* boson". Although the *W* boson was postulated as the weak analogue of the photon, it clearly had to have some uniquely different properties. All known weak processes such as beta decay involved an exchanged of electric charge between the currents J_μ and J'_μ. Thus the *W* boson had to exist in at least two charge states, W^+ and W^-. Most important, the *W* boson had to be extremely massive in order to explain the very short range of the weak force. In spite of these great differences between the *W* boson and the photon, the idea of *W* boson exchange was highly appealing because it suggested that the electromagnetic and weak interactions could be described by a common type of theory.

The interpretation of weak decays in terms of *W* boson exchange immediately encountered some difficulties. For example, it could not easily explain certain well known decays such as that of the

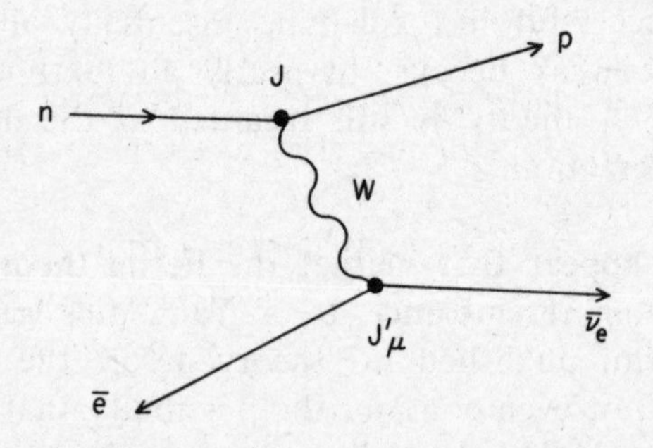

Fig. 6.2 Weak decay of the neutron $n \to p + e^- + \bar{\nu}_e$ in V–A theory is mediated by the exchange of a massive *W* boson between the neutron-proton current J_μ and the $e^- - \nu_e$ current J'_μ.

[7] N. Kemmer, *Phys. Rev.* **52**, 906 (1937).
[8] O. Klein, *Les Nouvelles Theories de la Physique*, 6 (Institut International de Cooperation Intellectuelle, 1938).

π-meson,

$$\pi^+ \longrightarrow \mu^+ + \nu_\mu \quad .$$

The decay of the π-meson obviously does not involve enough particles to be described by the same type of exchange process as the beta decay of the neutron. The π-meson decay process can be calculated in the V–A theory but the intuitively appealing picture of W boson exchange is lost.

6.3 Quarks and Weak Decays

The next important step toward a gauge theory not only solved some of the problems associated with the V–A theory but also involved a major breakthrough in the understanding of the strong interactions. During the decade following the Yang-Mills theory, the experimental study of elementary particles had entered the era of "big physics". Previously, observations of cosmic ray showers in emulsions and cloud chambers had revealed the existence of new unstable particles with properties very different from those of the known particles. The new high energy accelerators allowed physicists to produce these particles in the laboratory and study systematically the properties of the strong force by analysing tens of thousands of high energy interactions[9]. Scores of new mesons and baryons were soon discovered in these high energy scattering experiments. These particles appeared to be extremely short-lived excited states of the known nucleons and mesons and were called resonances. By studying the production and decay properties of these resonances, it was found that they could again be classified as eigenstates of isotopic spin and a new quantum number, called strangeness, which was also conserved by the strong interaction.

However, as more and more new particles were discovered, it became obvious that they could not all be considered as "elementary". Thus, Gell-Mann and others[10] proposed that the mesons and baryons are not elementary particles but are bound states of

[9] I. R. Kenyon, *Contem. Phys.* **13**, 75 (1972).
[10] M. Gell-Mann and Y. Ne'eman, *The Eightfold Way* (W. A. Benjamin, New York, 1964).

three new fundamental spin-$\frac{1}{2}$ constituents called "quarks".[b] The three quarks were defined to correspond to the old "up" and "down" eigenstates of isotopic-spin $\frac{1}{2}$ and to a new eigenstate of strangeness as shown in Fig. (6.3). These quark states, which are now called "flavors", formed the basis for the new SU(3) symmetry group which was a global symmetry like the old isotopic spin group. The quarks were postulated to have some very unusual properties. The most radical feature of the quarks was their electric charge, which was either $\frac{1}{3}$ or $\frac{2}{3}$ of the electron charge. The quark properties are shown in the table below:

Table 6.1: Properties of *u, d, s* Quarks

Flavor	I_3	Charge	Strangeness
u	$+\frac{1}{2}$	$+\frac{2}{3}$	0
d	$-\frac{1}{2}$	$-\frac{1}{3}$	0
s	0	$-\frac{1}{3}$	−1

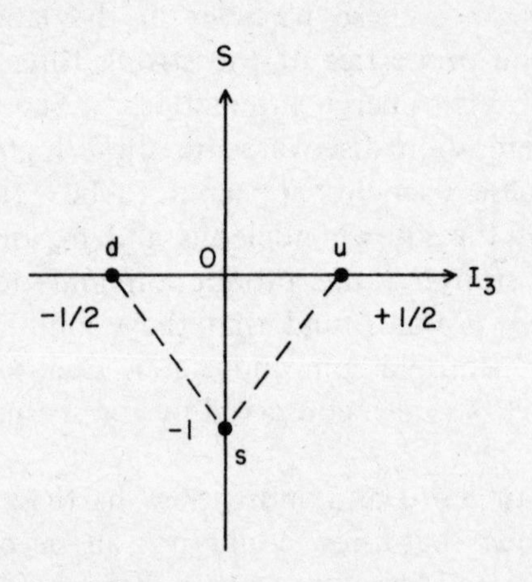

Fig. 6.3 Schematic representation of the *u, d* and *s* quark "flavors" as eigenstates of isotopic spin-$\frac{1}{2}$ and strangeness. Horizontal axis is the third component of isotopic spin I_3 and vertical axis is strangeness.

[b]The name quark is taken from a line in James Joyce's "Finnegan's Wake" which begins, "Three quarks for Muster Mark . . .".

Given the three quarks flavors above, the quantum numbers of all known strongly interacting particles could be understood by considering the baryons to be bound states of three quarks and the mesons to be bound states of quark and an antiquark. For example, the three charge states of the π-meson are given by the quark states

$$\pi^+ = u\bar{d} \quad ,$$

$$\pi^0 = \frac{1}{\sqrt{2}}(u\bar{u} - d\bar{d}) \quad ,$$

$$\pi^- = d\bar{u} \quad .$$

The proton is a *uud* quark state and the neutron is a *udd* state. The mesons and baryons can be grouped into eigenstates or "multiplets" of SU(3). The multiplets containing the meson and the proton and neutron are shown in Fig. (6.4). These two-dimensional multiplet diagrams have become part of the standard vocabulary of elementary particle physics. Further details about these diagrams can be found in the book by Gell-Mann and Ne'eman[10] or the very readable review article by Gasiorowicz and Rosner.[11]

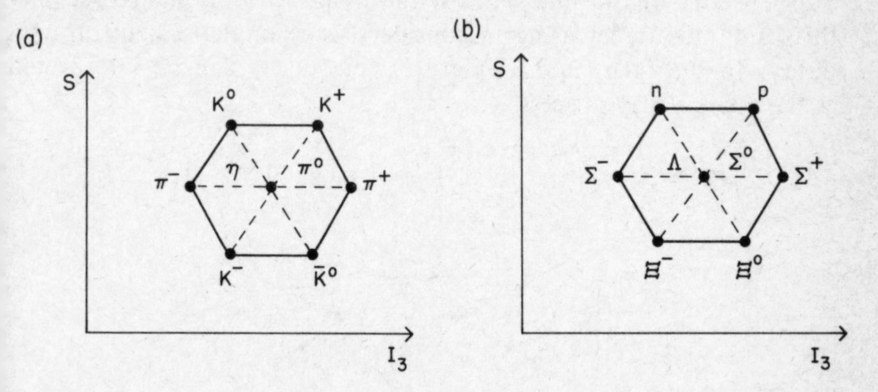

Fig. 6.4 Mesons and baryons classified in multiplets of SU(3) symmetry. (a) Octet of spin-0 mesons, (b) Octet of spin-$\frac{1}{2}$ baryons. Horizontal axis is the third component of isotopic spin I_3 and vertical axis is strangeness.

[11] S. Gasiorowicz and J. Rosner, *Am. Jour. Phys.* **49**, 54 (1981).

The success of the quark model completely changed the interpretation of the V–A weak theory. It became clear that the V–A theory should be applied to the quarks and not to the mesons and baryons. This new interpretation resolved the problem of the decay of the π meson. Since the π-meson contains a quark and antiquark, the decay can be understood as shown in Fig. (6.5). The u quark and the anti-d quark annihilate and emit a virtual W boson which subsequently decays to produce a μ^+ and ν_μ. The production of the W boson thus resembles the emission of a gamma ray from the annihilation of the e^+ and e^- in a positronium atom. This new quark model interpretation showed that the decay of the meson can be considered as a form of exchange process.

A much more important consequence of the quark model was that it revealed the underlying symmetry structure of the weak interaction and pointed out a deep connection between quarks and leptons. It had been suggested earlier by Schwinger[12] and Glashow[13] that the leptons must also carry isotopic spin like the strongly interacting particles. The quark model now provided the proper context for a unified treatment of the leptons and quarks as eigenstates of isotopic spin for the weak interaction. To see how this came about, let us again consider the example of neutron beta decay. In the quark model picture, beta decay can be interpreted as the quark decay process,

$$d \longrightarrow u + e^- + \bar{\nu}_e \quad ,$$

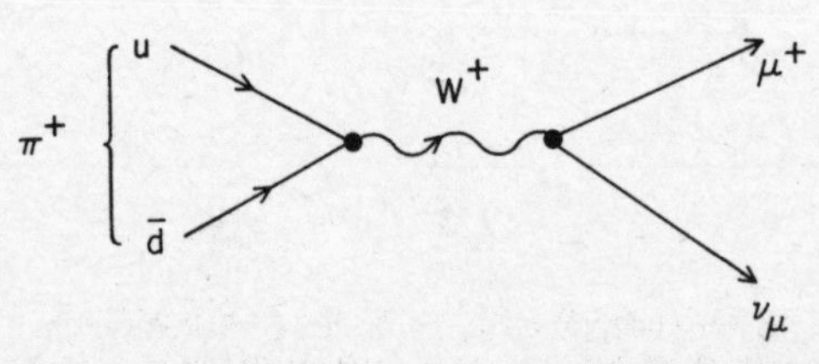

Fig. 6.5 Weak decay of a π-meson is explained in the quark model version of the V–A theory by the mutual annhilation of the u and $\bar{d}$ quarks in the π-meson. The emitted virtual W boson decays to a μ^+ and ν_μ.

[12] J. Schwinger, *J. Ann. Phys.* **2**, 407 (1957).
[13] S. Glashow, *Nucl. Phys.* **22**, 579 (1961).

as shown in Fig. (6.6). The other quarks in the neutron and proton are considered to act like spectators. The isotopic spin of the quarks changes by one unit in the transition $d \to u$. This gain of isotopic spin by the final quark can be understood if we now assume that the electron and its neutrino also are eigenstates of isotopic spin-$\frac{1}{2}$ like the u and d quarks. These states can be represented by column vectors in analogy with ordinary spin-$\frac{1}{2}$ states[c]

$$\begin{pmatrix} \nu_e \\ e^- \end{pmatrix}, \begin{pmatrix} u \\ d' \end{pmatrix} \quad . \tag{VI - 3}$$

We can then see that the W boson acts like a raising or lowering operator on the lepton and quark isotopic spin states. The isotopic spin lost by the initial quark is transferred by the W boson to the leptons. The fact that the electron is always accompanied by its neutrino then indicates that the isotopic spin is conserved. This interpretation also works for the decay of the π-meson.

The quark model led to a revolutionary and highly appealing

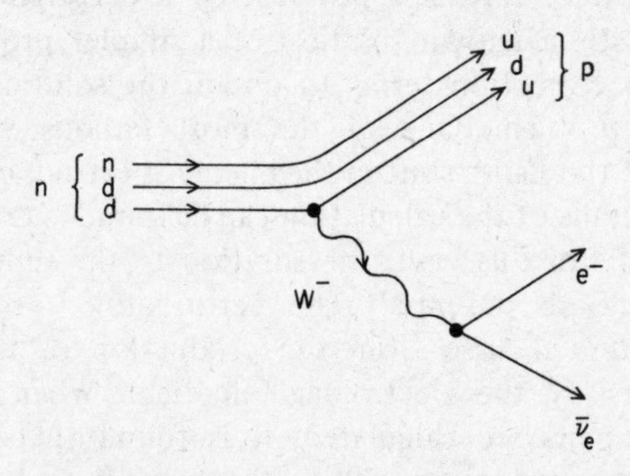

Fig. 6.6 Weak beta of the neutron $n \to p + e^- + \bar{\nu}_e$ is explained in the quark model version of the V–A theory by exchange of a W boson between the u and d quarks and the leptons e^- and $\bar{\nu}_e$. The other quarks in the neutron and proton act as spectators.

[c]The d′ quark in (VI - 3) is actually a mixture of a pure isotopic spin-$\frac{1}{2}$ d quark and the strange s quark. The s quark is also member of an isotopic spin-$\frac{1}{2}$ state, but its partner, the "charmed" quark, was not discovered until many years later.

picture of the weak interaction. Isotopic spin was now understood to be a quantum number associated with a symmetry of the weak interaction and not with the strong force. The most important effect of the quark model was that it revived the *W* boson. The identification of both quarks and leptons as eigenstates of weak isotopic spin made it clear that the *W* boson could act as a generator of the isotopic spin symmetry group as required by the original Yang-Mills theory. Thus it appeared that many of the necessary ingredients were available to formulate the weak interaction as a local gauge theory.

6.4 The "Dark Age" of Field Theory

Although conditions appeared to be right for a new attempt at a gauge theory, several theoretical difficulties blocked the way. The problem of the zero mass of the Yang-Mills gauge potential field was still unsolved. However there were even more severe problems with quantum field theory itself. These difficulties were associated with the well known technique of performing calculations by using "perturbation theory" in quantum electrodynamics (QED). The basic idea of solving a problem by a perturbative calculation is to start with a known solution of a simpler problem and then add on small correction terms to obtain the solution for the more complicated problem. One of the most famous examples is the calculation of the Lamb shift in the energy spectrum of the hydrogen atom. The details of the calculation can be found in texts on quantum field theory[14]. In this case, the solution to the simpler problem is just the Coulomb potential. The perturbative corrections to the Coulomb potential arise from the production of virtual electron-positron pairs by the electromagnetic field. When the effects of these virtual pairs are calculated, it is found that they can yield results for measurables quantities that appear to become infinite. One of the brilliant successes in the development of QED was the invention of a mathematical procedure known as "renormalization" which eliminated the infinities. The process of renormalization is

[14]J. D. Bjorken and S. D. Drell, *Relativistic Quantum Mechanics* (McGraw-Hill, New York, 1964).

possible only in a perturbative approach because it involves a delicate subtraction of one "infinite" term from another at each step in the perturbative calculation. The perturbative approach itself also depends crucially on the fact that the electromagnetic force is relatively weak. In QED, the strength of the interaction is measured by the coupling constant $\alpha = e^2/\hbar c = \frac{1}{137}$ which appears in Coulomb's law

$$\frac{F}{\hbar c} = \frac{\alpha Q Q'}{4\pi r^2} \quad , \qquad \text{(VI - 4)}$$

where Q and Q' are the charges measured in integer multiples of e. In perturbative calculations, higher-order contributions involve successively higher powers of α. Thus, the first-order correction is proportional to approximately 10^{-2}, the second order term to 10^{-4} and so forth.

Severe difficulties arose when the perturbative techniques which were so successful in QED were applied to the strong and weak interactions. It was immediately obvious that perturbative techniques could not work for the strong interaction. The corresponding coupling constant for the strong interaction

$$\frac{g^2}{\hbar c} \simeq 15 \qquad \text{(VI - 5)}$$

is greater than one. This meant that the higher-order perturbative corrections to a process such as meson exchange would not necessarily be smaller than the lower-order terms. Thus, it was impossible to estimate the accuracy of a calculation for a strong interaction from the order of the perturbative terms as one could do for QED.

A different type of problem arose in the calculation of weak processes. The weak Fermi coupling G in the V–A theory is even smaller than the electromagnetic, approximately 10^{-5}. Thus it appeared that the perturbative approach should work for calculations of weak interactions at least as well as in QED. However, the renormalization procedure broke down because new types of infinite terms appeared which could not be cancelled. An example of a scattering process which contains "non-renormalizable" infinite terms is the collision of a neutrino with an antineutrino. Although such an interaction may not be commonplace, except possibly in the interior of a star, it demonstrates the nature of the infinities. The lowest order contribution to neutrino-antineutrino scattering from perturbation theory is illustrated in Fig. (6.7a). The "cross-section", which measures the rate at which the reaction occurs has been calculated from the first-order term and is given by[15]

$$\sigma = \frac{GE^2}{12\pi} \quad , \tag{VI-6}$$

where E is the total energy of the reaction in the center-of-mass

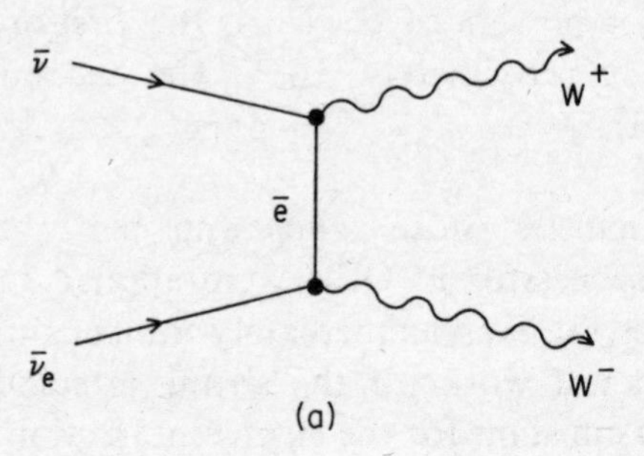

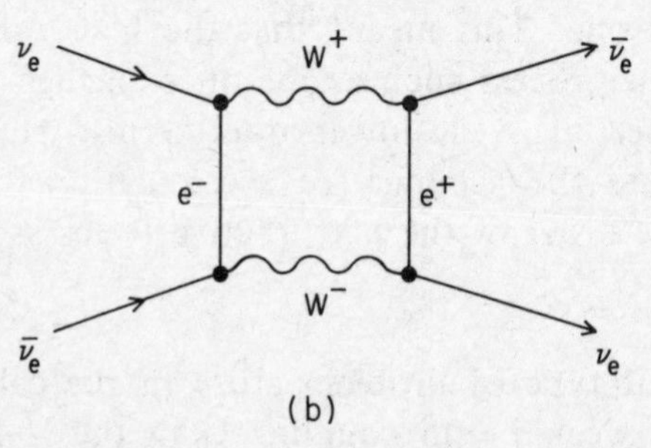

Fig. 6.7 Perturbative contributions to the scattering of a neutrino and an antineutrino in V–A theory. The lowest-order term (a) gives a cross-section which increases indefinitely with energy. The higher-order "loop" diagram (b) gives a "non-renormalizable" infinite contribution.

[15] M. Gell-Mann, M. L. Goldberger, N. Kroll, and F. E. Low, *Phys. Rev.* **179**, 1518 (1969).

system. This cross-section can become arbitrarily large as E increases. When the value of E is several hundred GeV, the cross-section eventually exceeds the maximum value allowed by ordinary quantum scattering theory, which is known as the "unitary limit".[16]

The possibility that infinities can arise in perturbative calculations had been pointed out much earlier by Heisenberg[17]. The accepted cure is to calculate higher-order perturbative terms which might conceivably damp out the undesirable growth of the cross-section. An example of a higher-order contribution is illustrated in Fig. (6.7b). This term is constructed by "sewing" together the lowest-order term with its mirror image such that the W^+ and W^- produced in Fig. (6.7a) now act as intermediate states in the scattering process. This contribution is known in the jargon of field theory as "box" or "loop" diagram for obvious reasons. Unfortunately, the calculation of this "loop" diagram gives an infinite result which cannot be cancelled by any other term in the V–A theory. The cure for these "non-renormalizable" infinities could not be found within the framework of standard quantum field theory and had to wait for the final rebirth of gauge theory.

As a result of the failure of perturbation theory and the presence of "non-renormalizable" infinities, the practice of quantum field theory entered a kind of "Dark Age". Despite the great success of the V–A theory and the quark model in explaining many weak and strong phenomena, it appeared that the enormously precise calculations possible in QED would never be duplicated for the nuclear interactions. The pessimistic tone of the period was expressed by Freeman Dyson in the quotation at the beginning of this chapter.

[16] D. H. Perkins, *Introduction to High Energy Physics* (Addison-Wesley, Reading, Mass., 1972). New. ed.
[17] W. Heisenberg, *Zeit. Physik* **101**, 533 (1936).

CHAPTER VII

THE BREAKING OF GAUGE SYMMETRY

Jean Buridan (AD 1300 – 1350)
S. Fubini, 1974[1]

7.1 Introduction

In the preceding chapters, we have discussed the properties of gauge theories which were always assumed to be exactly gauge invariant. However, we know from experience that gauge symmetry is more often broken than not in real physical systems. The original Yang-Mills gauge theory of isotopic spin was unsuitable for describing the strong nuclear force because it could not produce the necessary short-range interaction without explicitly violating gauge symmetry. This left electromagnetism as the only well understood gauge theory which agreed exactly with the known experimental facts. Thus, a major obstacle in the further development of gauge theories was the need to understand how a physical system can be described by a gauge theory and at the same time have properties which appear to violate gauge invariance.

The principle difficulty in understanding gauge symmetry breaking is that the most obvious symptoms are not directly related to any apparent source. In elementary particle physics, gauge symmetry breaking manifests itself through the short range of the interaction. The gauge field which mediates the interaction has a non-zero mass. The extremely short ranges of the weak and strong nuclear forces indicate that the masses involved are very large. Thus, in order to understand symmetry breaking for the fundamental

[1]S. Fubini, *Proc. Intl. Conf. on High Energy Physics* (Didcot, Rutherford Laboratory, England, 1974).

forces, one has also to understand how a gauge field with no mass can acquire a very large mass. The situation is further complicated by the fact that the internal quantum numbers such as isotopic spin are usually conserved by the interaction. This means that the Lagrangian describing the system is still manifestly gauge invariant even though the gauge field appears to violate gauge symmetry.

7.2 Symmetry Breaking of the Second Kind

In order to understand the origin of gauge symmetry breaking, we must re-examine the picture of gauge invariance presented in the preceding chapters to find the loophole that allows symmetry breaking. We will find that some of the essential clues to symmetry breaking originally came from solid state physics, a field that appears totally unrelated to the study of elementary particles. We will also see that the geometrical picture of gauge invariance again provides a simple and intuitive way to understand the general principles behind gauge symmetry breaking.

Let us approach the unknown problem of gauge symmetry breaking by first considering a well-known example of explicit symmetry breaking from atomic physics. We refer to the Zeeman effect in the hydrogen atom. In this case, the forces are electromagnetic and the symmetry involved is the 3-dimensional rotational symmetry of the Coulomb potential. This symmetry is broken by turning on an external magnetic field which selects out a particular direction in space. Before the symmetry breaking, the degeneracy of the atomic state described by the Schrödinger wavefunction can be seen in the arbitrary dependence on the azimuthal angle. After symmetry breaking, the dependence on the angle is no longer arbitrary and the degeneracy of the energy levels is lifted. The lesson to be learned here is that the obvious symptom of symmetry breaking, namely the splitting of energy levels, is actually related to a loss of part of the original symmetry. By introducing the external magnetic field, the symmetry is reduced from a complete rotational symmetry in three dimensions to the smaller set of rotations about the direction of the external magnetic field. This suggests a similar possibility in gauge theories.

As we saw in chapter III, the well-known arbitrariness in the value of the gauge vector potential can be interpreted in the geometrical picture as an arbitrariness in the choice of a direction in the internal symmetry space. This suggests that there may be an analogy between the spatial rotations in the Zeeman effect and the internal rotations in a gauge theory. Pursuing this analogy further, we are then motivated to break the gauge symmetry by introducing some external constraint which will limit the arbitrariness of the internal-space direction of the gauge field itself. However, constraining the gauge field is a non-trivial problem since we also want to preserve the gauge invariance of the Lagrangian at the same time. Thus, we cannot simply introduce an external field into the Lagrangian that fixes the internal-space direction to be the same at all points. This would lead us to the same type of problems that we encountered with the mass term in the Yang-Mills theory. We shall see that the best way to solve this problem is to take a different approach. Rather than starting with the solution for the unbroken symmetry and trying to modify it by including *ad hoc* symmetry breaking terms, we will try to construct the solution for broken gauge symmetry by using very general physical arguments.

7.3 Geometry of Symmetry Breaking

Since we are not allowed to put in an explicit symmetry breaking term, what other alternatives are there? The answer must be that the symmetry breaking is "induced" through the normal dynamical interaction of the gauge field with an external charged particle or system. In the following discussion, we will therefore re-examine our basic ideas about the minimal coupling of the gauge field and see how they can be generalized to include symmetry breaking. As before, we will see that we can use the simple geometrical picture instead of complicated mathematics for discussing symmetry breaking.

For a point-like free particle, the change in the internal direction between different space-time points is completely determined by the minimal coupling to the gauge potential field. The internal direction is free to rotate in any way required by the gauge field.

Thus the interaction of a gauge field with a free particle, or even an ensemble of free particles, does not impose any constraints on the possible orientations of the internal direction which might lead to symmetry breaking. In order for a particle to break the symmetry imposed by its interaction with the gauge field, the particle must somehow determine its own direction in the internal symmetry space independently of the requirement of the external gauge field. At the same time, the particle must still be minimally coupled to the gauge field to preserve the invariance of the Lagrangian. In less geometrical language, this means that the wavefunction must have an intrinsic relative phase factor which is not the same function of space-time position x as the phase change required by the usual minimal coupling of a free particle with the gauge potential. Clearly, these requirements mean that we are no longer discussing an ordinary point-like particle but a new type of physical system which interacts with the gauge field. We will refer to such systems as "self-coherent" for reasons which will become clear soon. In the geometrical picture, the phase of a self-coherent system will trace out a path in the internal space that is different from that of a free point particle in the same gauge field, as illustrated in Fig. (7.1).

Although it may appear that the phases of the self-coherent system and the external gauge field must be completely different

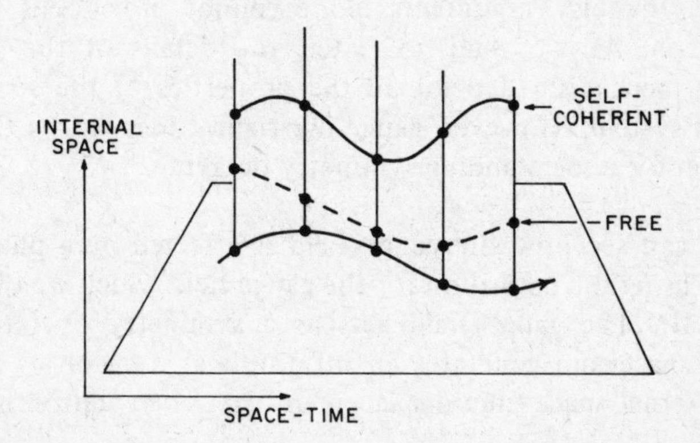

Fig. 7.1 Behaviour of the phases of a free test particle and a self-coherent system in the internal symmetry space. The phase of the free particle is determined by the external gauge field.

in order to break the gauge symmetry, this is not the case. The phase relationships must still be consistent at every point in space-time. The reason for this is that the interaction is still given by the minimal coupling or gauge covariant-derivative. We saw in chapter III that the minimal coupling form is a straightforward consequence of the basic postulates of local gauge invariance in the geometrical picture; thus, the minimal coupling form must be maintained in order to preserve the manifest gauge invariance of the system, in particular the Lagrangian. These requirements clearly lead to a potential conflict between the gauge field and a self-coherent system. The gauge field will try to rotate the local phase of the self-coherent system at each point just as it would the phase of a free test particle. However the rotation will not be allowed because of the intrinsic phase relation of the system. The resolution of this apparent difficulty is quite simple. The gauge field itself must undergo some change in order to adjust its phase requirements to that of the self-coherent system. This is the actual meaning of gauge symmetry breaking. The gauge field is "broken" from the free particle form to the "self-coherent" form.

An important omission in the above argument is that we have not specified any dynamical mechanism which describes just how the gauge field is broken or adjusted to its new phase relation. General symmetry arguments alone cannot provide us with that information. As we shall see later, the details of the symmetry breaking mechanism depend on the properties of the specific self-coherent system. However, gauge invariance does tell us the general formalism for understanding symmetry breaking.

We can see how the gauge field is affected by a phase adjustment by using the definition for the gauge field which we obtained in chapter III. The gauge field acts as a symmetry operator in the internal space and generates an infinitesimal rotation as the result of an external space-time displacement $\mathrm{d}x$. The rotation is given by

$$\delta\theta = qA_\mu \mathrm{d}x^\mu \quad , \qquad \text{(VII - 1)}$$

where we have previously defined the gauge field as

$$A_\mu = \sum_k A_\mu^k(x) F_k \quad , \tag{VII - 2a}$$

$$A_\mu^k(x) = \partial_\mu \theta^k(x) \quad . \tag{VII - 2b}$$

The generators F_k are not changed by the symmetry breaking since they define the gauge symmetry group. As we argued above, only the phase or internal rotation angles $\theta^k(x)$ are adjusted by the interaction with the self-coherent system. However, the internal angles provide the only functional dependence of the internal space on the space-time coordinate x. Thus, the breaking of gauge symmetry by a phase adjustment must produce a change in the space-time dependence of the gauge field itself. The amount of this change must be such that the local phase required by the broken gauge field is then consistent with that of the self-coherent system at every point in space-time.

We can use a simple geometrical argument to deduce the general form of the phase-adjusted gauge field after symmetry breaking. Consider the example shown in Fig. (7.2). At the position x, we assume that a free particle and a self-coherent system have the same initial orientation in the internal space as indicated by the direction of the vectors. After an infinitesimal displacement dx, the free

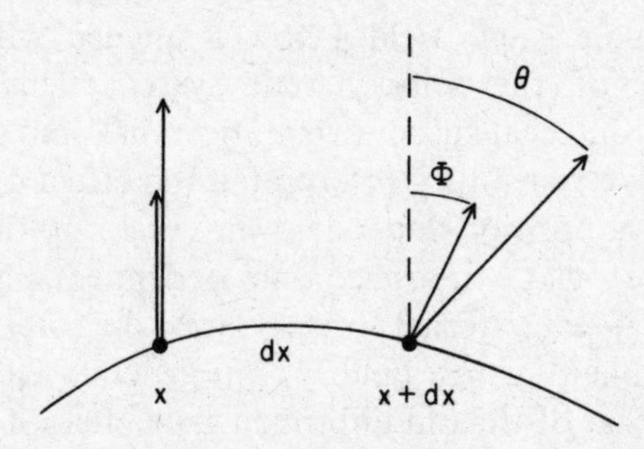

Fig. 7.2 The difference in the rotations of the internal-space directions of a free test particle and a self-coherent system as the result of an infinitesimal space-time displacement from x to x + dx.

particle phase has been rotated through a small angle $\delta\theta$ by the external gauge field. However, the change in phase of the self-coherent system between x and x + dx is given by a different angle $\delta\Phi$. We define the difference between these two phase changes to be

$$\delta\lambda = \delta\theta - \delta\Phi \quad . \tag{VII - 3}$$

The phase changes can be expressed in terms of gauge potentials by using (VII-1). To first order in dx, we can write

$$\delta\theta = qA_\mu dx^\mu \quad , \tag{VII - 4a}$$

$$\delta\lambda = qW'_\mu dx^\mu \quad , \tag{VII - 4b}$$

$$\delta\Phi = \partial_\mu \Phi dx^\mu \quad , \tag{VII - 4c}$$

where W'_μ is the new "phase-adjusted" gauge field after symmetry breaking. We then obtain from Eq. (VII-3),

$$W'_\mu = A_\mu - \frac{1}{q}\partial_\mu \Phi \quad . \tag{VII - 5}$$

We see from this relation that the new gauge field W'_μ acting on a free particle, would generate an internal rotation which agrees exactly with the phase change of the self-coherent system. From the point of view of the free test particle (i.e. physicist inside the elevator), it appears that the local frame which was formerly aligned with the unbroken gauge field A now is aligned with the internal-space direction of the self-coherent system. This rather formal change in the internal-space orientation has important physical consequences. We previously interpreted the effect of a real external gauge field as a position-dependent internal rotation; conversely, we must now associate the space-time dependent phase adjustment with the effect of real external forces. Thus, the force field produced by the new "broken" gauge field W'_μ, in general, will be different in character from that of the old unbroken gauge field A_μ.

Since the phase change in the gauge field need only be local, the space-time properties of the gauge field may be modified in a relatively small spatial region. This depends on the detailed nature of the self-coherent system. We will see later that this provides a simple explanation for the most familiar symptoms of symmetry breaking such as the apparent short-range behavior of a normally long-range gauge field.

7.4 The Broken Gauge Superconductor

We will now examine a real physical system in which gauge symmetry is broken in order to see how the special self-coherent phase relation is actually realized by dynamical mechanisms. The example we will discuss is the familiar phenomenon of superconductivity. The superconductor is particularly well suited for our purposes because only electromagnetism is involved and a simple nonrelativistic semiclassical model for superconductivity already exists.

The unique coherent properties of superconductors have been well known for some time. Nearly thirty years ago, Fritz London remarked that "superconductivity is a remarkable manifestation of quantum mechanics on a truly macroscopic scale".[2] He pointed out that many of the salient features of superconductivity, such as the Meissner effect and flux quantization, can be understood by describing the electrons in the superconductor as a single coherent system which behaves like a free quantum-mechanical particle. This macroscopic character of the current results from a coherent superposition of many electron wavefunctions in the superconductor. As we shall see, this electron current is the "self-coherent" system which is responsible for the gauge symmetry breaking.

The rigorous derivation of how the electrons form a coherent macroscopic system was first given by Bardeen, Cooper and Schrieffer[3] using quantum field theory techniques. However, for our purposes, a qualitative semiclassical description is sufficient[4]. Roughly speaking, the interaction of the conduction electrons with the atomic

[2]F. London, *Superfluids* Vol. 1 (Wiley, New York, 1950).
[3]J. Bardeen, L. N. Cooper and J. R. Schrieffer, *Phys. Rev.* **108**, 1125 (1957).
[4]V. Weisskopf, *Contem. Phys.* **22**, 375 (1981).

lattice of the superconductor produces an attractive force between the electrons. When the electron energies are sufficiently small, this attractive force overcomes the normal Coulomb repulsion and binds electrons together into Cooper pairs. The electron spins are oriented in opposite directions so that a Cooper pair acts like a spin-0 boson with two units of negative charge. Owing to the extremely weak binding, the effective size of a Cooper pair is very large about 10^{-4} cm. Every Cooper pair overlaps with approximately 10^6 other Cooper pairs, and this superposition results in a strong correlation which "locks" the phases of the wavefunctions together coherently over macroscopic distances. The current in the superconductor therefore behaves like a single free quantum-mechanical particle.

The interaction of an external magnetic field with the coherent electrons in the superconductor results in the Meissner effect. When the magnetic field penetrates into the superconductor, it induces a flow of Cooper pair current. This current generates its own magnetic field which cancels the external field. However, the cancellation is not complete; the external magnetic field penetrates a small distance into the superconductor, its strength decreasing exponentially with depth. This short-range behaviour of the field can be easily obtained using simple quantum-mechanical arguments. The coherent system of Cooper pair electrons can be described by the Schrödinger wavefunction

$$\Psi(\mathbf{x}, t) = \left(\frac{N}{2}\right)^{1/2} \exp[-2ie\Phi(\mathbf{x}, t)/\hbar c] \quad , \qquad \text{(VII - 6)}$$

where $N/2$ is the density of Cooper pairs which we assume to be constant. The current density is obtained from

$$\mathbf{J} = -\frac{e\hbar}{2im}(\bar{\Psi}\boldsymbol{\nabla}\Psi - \Psi\boldsymbol{\nabla}\bar{\Psi}) - \frac{2e^2}{mc}|\Psi|^2\mathbf{A} \quad . \qquad \text{(VII - 7)}$$

Substituting (VII-6) into the current gives

$$\mathbf{J} = \frac{Ne^2}{mc}(\boldsymbol{\nabla}\Phi - \mathbf{A}) \quad . \qquad \text{(VII - 8)}$$

It is customary[5] to use the Coulomb gauge, $\nabla \cdot \mathbf{A} = 0$, to show that conservation of the current requires $\nabla^2 \Phi$ to vanish. This implies that $\nabla\Phi$ is constant. Choosing $\nabla\Phi$ to be zero then leads to the familiar London equation,

$$\mathbf{J} = -\frac{Ne^2}{mc}\mathbf{A} \quad . \tag{VII - 9}$$

Combining the London equation with Maxwell's equation yields a relation for the magnetic field in the superconductor

$$\nabla^2 \mathbf{B} = -\frac{Ne^2}{mc}\mathbf{B} \quad . \tag{VII - 10}$$

The solution to this equation is an exponential which shows that the magnetic field decreases rapidly inside the superconductor[a]. Thus the Meissner effect generates a short-range field without violating the gauge invariance of Maxwell's equations.

How does this example of the superconductor help us to understand how a general gauge field acquires a non-zero mass? The essential clue is provided by the choice of the phase condition, $\nabla\Phi = 0$, which leads to the London equation. Let us see how this condition is related to the general "broken" gauge field W'_μ given in (VII - 5). We first observe that the London equation can be obtained directly from the current (VII-7) without choosing the Coulomb gauge if we simply replace the original potential A_μ by the new potential W'_μ. The potential W'_μ is precisely the solution obtained by specifying the phase condition, $\nabla\Phi = 0$. How can this be? From the derivation of W'_μ, we see that it is clearly defined to generate a phase change equal to the difference between the phase change of a free particle in the potential A_μ and the phase change of the Cooper pair current. The potential W'_μ therefore "interpolates" between the two extreme limits corresponding to the external field A_μ and the field allowed deep inside the superconduc-

[a]The penetration depth of the magnetic field is given by $\lambda = [mc^2/4\pi e^2 N]^{1/2}$, which is called the London penetration depth. The value of λ is typically 30 – 50 nanometers.

[5]R. P. Feynman, R. B. Leighton and M. Sands, *Feynman Lectures in Physics*, Vol. **3** (Addison-Wesley, Reading, Mass., 1965).

tor. We have seen that the short-range magnetic field in the Meissner effect interpolates in space between the external magnetic field and zero internal field. It is now clear that the "broken" gauge potential W'_μ interpolates the phase between the external and internal values.

It is clear from our discussion of the Meissner effect that symmetry breaking occurs because the Cooper pair current violates gauge invariance. The reason for this peculiar property of the Cooper pair current is the coherent phase of the wavefunction. The functional dependence of the phase Φ on x is determined only by the binding and overlap of the Cooper pairs, which are independent of any external gauge fields. Thus, the Cooper pair current is the self-coherent system which is necessary to break the gauge symmetry. It is important to note, however, that the gauge invariance is still preserved by the local interaction between the magnetic field and the Cooper pair current. The form of the interaction is just given by the "minimal" electromagnetic coupling at each point x which is already built into the current (VII-7). This guarantees that the Lagrangian and the Maxwell equations are still manifestly gauge invariant.

7.5 Spontaneous Symmetry Breaking

The ingenious symmetry breaking property of the superconductor was generalized for non-Abelian gauge theories by Higgs[6], Kibble[7] and others[8,9]. This mechanism was renamed "spontaneous symmetry breaking" by Baker and Glashow[10] because it did not require any explicit mass term in the Lagrangian to manifestly violate gauge invariance.

In place of the Cooper pairs, one postulates the existence of a new fundamental spin-0 field ϕ, which is called a Higgs field. The Higgs field is uncharged but it may carry quantum numbers like weak isotopic spin. The interaction of the Higgs field with an

[6]P. W. B. Higgs, *Phys. Lett.* **12**, 132 (1964), **13**, 508 (1964); *Phys. Rev.* **145**, 1156 (1966).
[7]T. W. B. Kibble, *Phys. Rev. Lett* **13**, 585 (1964).
[8]P. W. Anderson, *Phys. Rev.* **130**, 439 (1963).
[9]F. Englert and R. Brout, *Phys. Rev. Lett.* **13**, 321 (1964).
[10]M. Baker and S. L. Glashow, *Phys. Rev.* **128**, 2462 (1962).

external gauge field is described by the following Lagrangian,

$$\mathcal{L} = \tfrac{1}{2}|D_\mu\phi|^2 - \tfrac{1}{4}|F_{\mu\nu}|^2 \quad . \qquad \text{(VII - 11)}$$

The first term is the kinetic energy of the Higgs field and the second term gives the energy density of the gauge field. Using this Lagrangian in the Euler-Lagrange equations will yield the Yang-Mills equations again and the Klein-Gordon equation for a scalar field.

In order for the Higgs field to play the same role as the Cooper pair current, it is assumed that the potential energy of the Higgs field is given by a special function

$$V(\phi) = \mu^2|\phi|^2 + \lambda(|\phi|^2)^2 \quad . \qquad \text{(VII - 12)}$$

The form of $V(\phi)$ was first proposed before the BCS theory by Ginzburg and Landau[11] as a purely phenomenological description of the free energy density of the superconductor. In gauge theory, the potential is interpreted as a kind of "self-interaction" of the Higgs field. When viewed from the internal symmetry space, this potential has several unusual properties. As we noted in chapter II, the internal space associated with the U(1) gauge group is a ring of phase values. Since the potential $V(\phi)$ is invariant under phase rotations of U(1), it also has a circular "cross section" in the internal space. This is illustrated in Fig. (7.3), where the vertical axis measures the potential energy. What is most interesting about $V(\phi)$ is that the shape of the potential changes dramatically when the sign of the parameter μ^2 is changed. This leads to a degenerate ground state of ϕ for negative value of μ^2. We can easily see this by finding the minima of $V(\phi)$. For positive μ^2, there is a unique minimum when $\phi_0 = 0$. For negative values of μ^2, the minimum of $V(\phi)$ yields a non-vanishing ground state given by

$$\phi_0 = \left(-\frac{\mu^2}{\lambda}\right)^{1/2} \exp(i\theta(x)) \quad , \qquad \text{(VII - 13)}$$

where $\theta(x)$ is the phase or internal angle. Due to the gauge invariance of $V(\phi)$, an arbitrary gauge transformation can rotate this ground

[11] V. Ginzburg and L. D. Landau, *Zh. Eksp. Teor. Fiz.* **20**, 1064 (1950).

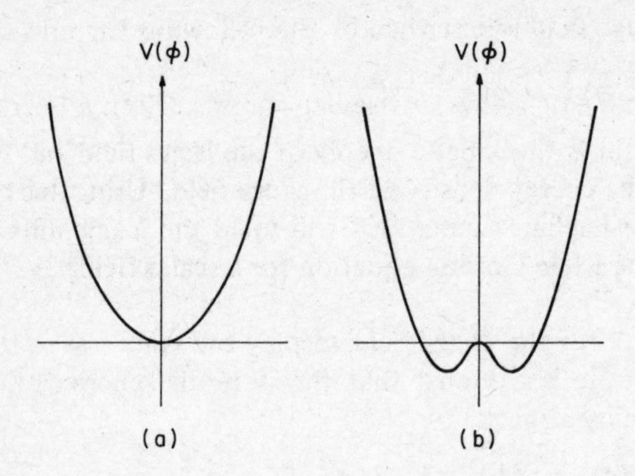

Fig. 7.3 The variation of the shape of the self-interaction potential $V(\phi)$ of the scalar Higgs field. A single minimum occurs for (a) $\mu^2 > 0$ while an infinite number of degenerate minima occur for (b) $\mu^2 < 0$.

state into an infinity of equivalent ground states,

$$\phi_0 \longrightarrow \phi_0' = \phi_0 \exp(i\theta') \quad , \tag{VII - 14}$$

all of which are equally good minimum solutions of $V(\phi)$.

In the Ginzburg-Landau model, the change in shape of the potential $V(\phi)$ is interpreted physically as a phase transition between the superconducting and normally-conducting states. The sign of the parameter μ^2 depends on whether the temperature is above or below the critical temperature T_c at which superconductivity occurs. When the temperature is above T_c, the unique minimum $\phi_0 = 0$ indicates that the electrons are free particles and do not form Cooper pairs. Below the critical temperature, the Cooper pairs condense into a ground state whose phase is degenerate. We note that this demonstrates a loss of symmetry just as we discussed earlier in the example of the Zeeman effect. In the normally-conducting phase, the electrons are free particles so that their phases are uncorrelated and can be arbitrarily rotated at all points in space. In the superconducting phase, the individual rotational freedom is lost because the electrons form a coherent system. However, the coherent system can still be rotated as a whole.

The degenerate Higgs ground state ϕ_0 plays the role of the self-coherent field which breaks the gauge symmetry. The degeneracy of the ground state takes on special significance in a quantum field theory because the ground state of a system defines the "vacuum". This vacuum is required to be unique. The phase of the vacuum cannot be arbitrary at each point in space-time like that of a free particle. Once a particular value of the phase is chosen in order to uniquely define the vacuum state, the value remains the same at all points. This means that the vacuum is not invariant under arbitrary gauge transformation like (VII-14). The Higgs vacuum therefore breaks the gauge symmetry like the Cooper pair current in a superconductor. An amusing analogy was proposed by A. Salam[12] to illustrate the symmetry breaking by the Higgs vacuum:

> *Imagine a banquet where guests sit at round tables. A bird's eye view of the scene presents total symmetry, with serviettes alternating with people around each table. A person could equally well take a serviette from his (or her) right or his left. The symmetry is broken when one guest decides to pick up from his left and every one else follows suit.*

To see how the Higgs ground state breaks the gauge symmetry, we only need to look at the kinetic energy term of the Lagrangian

$$\text{K.E.} = \tfrac{1}{2}|D_\mu \phi|^2 \quad , \tag{VII - 15}$$

which contains the minimal coupling interaction between the gauge field and the degenerate ground state. From (VII-5), we see that the kinetic energy can be rewritten in terms of the broken potential W'_μ as

$$|D_\mu \phi_0|^2 = q^2 |W'_\mu|^2 |\phi_0|^2 \quad , \tag{VII - 16}$$

which has just the desired form for a gauge field mass term. We can easily read off the mass value to be

$$m = q\sqrt{|\phi_0|^2} = q\left(-\frac{\mu^2}{\lambda}\right)^{1/2} . \tag{VII - 17}$$

[12] A. Salam, *CERN Courier* 17, 271 (1977).

Has some sort of miracle occured? We have performed some apparently trivial mathematical manipulations and obtained a mass term for the gauge field. Although ingenious, the spontaneous symmetry breaking is not quite a miracle. The mass term was contained in the kinetic energy of the Higgs ground state from the beginning and was simply hidden. It is important to note that the mass term only arises from the interaction with the ground state. If the scalar field ϕ in (VII-15) is not the ground state, then the phase angle is not coherent and one obtains just the ordinary kinetic energy of a free particle.

The phase adjustment of the broken gauge field W'_μ is also considered to be a new generalized form of gauge transformation which is called the "unitarity gauge transformation".[13] This is not surprising since the definition (VII-5) of W'_μ closely resembles a gauge transformation. However, for an ordinary gauge transformation of the form

$$U = \exp(i\lambda) \quad , \tag{VII - 18}$$

the function λ can be arbitrarily chosen, whereas the phases that we are considering depend on the detailed dynamics of the self-coherent system. In addition, it is usually assumed that a gauge transformation does not alter the physical space-time properties of a system. Thus we have chosen to distinguish the phase adjustment from an ordinary gauge transformation for the sake of pedagogical clarity.

7.6 Goldstone's Theorem

The preceding discussion of the superconductor and the Higgs vacuum touches upon an important theorem by Goldstone[14,15] which was significant in the early understanding of gauge symmetry breaking. The theorem states that if a theory has an exact symmetry, such as a gauge symmetry, which is not a symmetry of the vacuum, then the theory must contain a zero-mass particle. In the case of the

[13] S. Weinberg, *Phys. Rev.* D7, 1068 (1973).
[14] J. Goldstone, *Nuovo Cimento* **19**, 154 (1961).
[15] J. Goldstone, A. Salam and S. Weinberg, *Phys. Rev.* **127**, 965 (1962).

Higgs field, Goldstone's theorem implies that there is massless scalar field other than ϕ itself. In order to see this, we turn off the electromagnetic field and look at the kinetic energy term (VII-15). Instead of substituting the broken field W'_μ, we will use the ground state given in (VII - 14) and calculate the derivative to obtain

$$D_\mu \phi_0 = i\partial_\mu \theta \phi_0 \quad . \qquad \text{(VII - 19)}$$

If we interpret the angle $\theta(x)$ as a new scalar field, it can be seen that (VII-19) will give rise to a Klein-Gordon equation for a zero-mass particle. The physical meaning of this massless field is that it is a phase or "angular" wave that can propagate through the ground state medium. This wave is not just a mathematical fiction. Such waves actually occur in condensed matter systems. In the superconductor, these waves are known as plasma oscillations. For a ferromagnet, the waves are called Bloch spin waves or magnons, and correspond to a precession of the direction of magnetization. The importance of the Goldstone waves for a broken gauge symmetry is that they are absorbed by the external gauge field as shown in (VII-16). The presence of the Goldstone waves was undesirable in elementary particle physics because they represent massless scalar particles which have never been observed. Thus the Higgs mechanism showed how the Goldstone waves could be eliminated.

7.7 A Parable about Symmetry Breaking

The picture at the beginning of this chapter illustrates a parable by the French philosopher Jean Buridan [A.D. 1300 – 1350] which has now become a part of the literature of gauge theory[1]. The donkey is placed mid-way between two identical amounts of food and cannot decide which one to eat. The indecision of the donkey is gauge invariance. What is needed to prevent the donkey from starving is an external influence – the self-coherent Higgs field – to make the choice for the donkey.

CHAPTER VIII

THE WEINBERG-SALAM UNIFIED THEORY

What God hath put asunder no man shall ever join.

W. Pauli[1]

8.1 Introduction

On the tenth anniversary of the Yang-Mills theory, there was still no successful gauge theory of the nuclear forces. A significant breakthrough had been achieved, as we saw in chapter VII, by finally solving the problem of the zero gauge field mass through spontaneous symmetry breaking. And the weak interaction, not the strong force, appeared to be the best candidate for a gauge theory. Yet the insurmountable difficulty of the "non-renormalizable" infinities still plagued both the weak interaction and quantum field theory. Nevertheless, the final step toward a successful gauge theory was taken almost simultaneously by Steven Weinberg[2] at MIT and Abdus Salam[3] at Imperial College, London. They boldly ignored the problem of the "non-renormalizable" infinities and instead proposed a far more ambitious unified gauge theory of the electromagnetic and weak interactions.

The idea of unifying the weak and electromagnetic interactions into a single gauge theory did not originate with Weinberg and Salam. It had been suggested much earlier by Schwinger[4] and Glashow[5]. As we noted in chapter VI, Glashow and Schwinger

[1] "Folklore"
[2] S. Weinberg, *Phys. Rev. Lett.* **19**, 1264 (1967).
[3] A. Salam, *Proc. of the 8th Nobel Symp. on Elementary Particle Theory*, ed. N. Svartholm (Almquist Forlag, 1968).
[4] J. Schwinger, *Ann. Phys.* **2**, 407 (1956).
[5] S. L Glashow, Ph. D. Thesis (Harvard University, 1958).

had also pointed out that the leptons should carry weak isotopic spin like the mesons and baryons involved in the weak decays. Thus much of the detective work had already been done in untangling the weak interactions and laying the logical foundation for a gauge theory. However, the essential ingredient missing in all of these earlier theoretical attempts was an understanding of the origin of the gauge field masses. Weinberg and Salam were the first to realize that the Higgs mechanism could supply the last piece of the puzzle for a unified gauge theory.

8.2 Weak Interactions Revisited

In chapter VI, we noted that the V−A theory of the weak interaction contained most of the essential ingredients necessary for a Yang-Mills gauge theory. We can now set up the mathematical machinery to formulate the weak interaction as a gauge theory with a spontaneously broken symmetry. For the sake of simplicity, we will begin by considering only leptons since the differences between leptons and quarks are not important for our discussion.

The basic interaction of a lepton with a vector potential field is given by the familiar "canonical momentum" or minimal coupling. The kinetic energy term in the Lagrangian can be written

$$\text{K.E.} = \bar{\psi}\gamma^{\mu} D_{\mu} \psi \qquad \text{(VIII - 1)}$$

where ψ is a 4-component Dirac spinor and γ_{μ} are the usual 4×4 Dirac matrices. The kinetic energy can be separated into a purely "kinetic" term and an interaction term as follows

$$\begin{aligned} \text{K.E.} &= \bar{\psi}\gamma^{\mu}\partial_{\mu}\psi - iq\bar{\psi}\gamma^{\mu}A_{\mu}\psi \\ &= \bar{\psi}\gamma^{\mu}\partial_{\mu}\psi - iqj^{\mu}A_{\mu} \quad . \end{aligned} \qquad \text{(VIII - 2)}$$

This form shows us explicitly how the lepton current j^{μ} couples to the vector potential A_{μ}. When A_{μ} is the electromagnetic potential, the corresponding current j^{μ} is given by

$$j^{\mu}_{\text{em}} = \bar{\psi}\gamma^{\mu}\psi \quad . \qquad \text{(VIII - 3)}$$

For the weak interaction potential W_μ, the charged weak current consists of the electron and its neutrino. This current is written

$$j^\mu_{\text{wk}} = \bar{\nu}\gamma^\mu(1-\gamma_5)e \qquad \text{(VIII - 4)}$$

where the particle symbols $\bar{\nu}$ and e represent the appropriate Dirac spinors. The contributions from the μ and τ leptons are treated in the same way. The known parity violating property of the weak interactions is built in through the factor $(1-\gamma_5)$, which projects out only the left-handed spin polarization state of the electron and neutrino. The presence of the γ_5 matrix means that the weak current is not a pure vector like the electromagnetic current (VIII-3) but a combination of vector and axial-vector contributions.

The first step towards a modern gauge version of the weak theory is to realize that the known lepton pairs (ν_e, e), (ν_μ, μ) or (ν_τ, τ) must be considered as $I = \frac{1}{2}$ spinor states or "doublets" of isotopic spin like the neutron and proton. The charged weak current can then be written in the form

$$j^\mu_{\text{wk}} = \bar{\psi}\tau_\pm\gamma^\mu(1-\gamma_5)\psi \qquad \text{(VIII - 5)}$$

where ψ represents the isotopic spinor

$$\psi = \begin{pmatrix} \nu_e \\ e \end{pmatrix} . \qquad \text{(VIII - 6)}$$

The $\tau_\pm$ are the Pauli matrices

$$\tau_+ = \begin{pmatrix} 0 & 1 \\ 0 & 0 \end{pmatrix}, \qquad \tau_- = \begin{pmatrix} 0 & 0 \\ 1 & 0 \end{pmatrix}, \qquad \text{(VIII - 7)}$$

which raise and lower the isotopic spin, respectively.

Since the electromagnetic current is purely vector while the weak current has a V–A structure, it is also convenient for our later discussion to adopt the convention of explicitly separating the left and right-handed spin polarization states of the spinors. We define the left-handed isotopic-spin doublet

$$L = \begin{pmatrix} \nu_e \\ e \end{pmatrix}_L = \tfrac{1}{2}(1 - \gamma_5) \begin{pmatrix} \nu_e \\ e \end{pmatrix} \quad . \tag{VIII - 8}$$

The neutrino is massless and is always left-handed by convention; thus, the right-handed state contains only the electron,

$$R = e_R = \tfrac{1}{2}(1 + \gamma_5)e \quad . \tag{VIII - 9}$$

Thus from the point of view of the weak interaction, the electron has a dual classification. The right-handed electron state is a weak isotopic-spin singlet state ($I = 0$) while the left-handed electron state is part of an isotopic-spin doublet with the neutrino.

The next step toward a gauge theory is to describe the electromagnetic and weak interactions of the leptons with the same type of minimal coupling to the gauge fields. Thus we will write the leptonic kinetic energy term of the Lagrangian as

$$\begin{aligned} \bar{\psi}\gamma^\mu D_\mu \psi = {} & \bar{R}\gamma^\mu(\partial_\mu - \mathrm{i}qA_\mu)R \\ & + \bar{L}\gamma^\mu(\partial_\mu - \mathrm{i}qA_\mu - \mathrm{i}gW^k_\mu \tau_k)L \quad , \end{aligned} \tag{VIII - 10}$$

where q is the usual electric charge. We have introduced a new weak isotopic "charge" g which is related to the phenomenological Fermi coupling constant G in V–A theory by

$$\mathrm{G} = \frac{g^2}{2M_W^2} \quad , \tag{VIII - 11}$$

where M_W is the mass of the W boson. It is easy to see that a complete Yang-Mills theory for the weak and electromagnetic interactions can be constructed by adding the appropriate terms to (VIII - 10) for the energy densities of the A_μ and W_μ fields and the lepton masses.

However, the theory would not be unified and the W_μ field would be massless since the weak isotopic-spin gauge symmetry is not broken. We will consider these two problems separately in the following discussion.

8.3 Unification without Renormalization

In order to upgrade the old V–A theory into a unified gauge theory, we need a third component of the gauge field corresponding to the isotopic-spin operator τ_3. This new field W_μ^3, which carries no electric charge, will complete the set of generators for an SU(2) weak isotopic-spin group. But how do we associate this third component with a physical gauge field? It might seem that an attractive choice for this third component is the electromagnetic potential A_μ itself. This would lead to a very economical unification within a single SU(2) gauge group. This approach was tried by Glashow[6] and others, but it was unsuccessful. The structure of the SU(2) group was too restrictive to accomodate a conserved electric current and the two charged W boson fields with the correct coupling strengths. As stated by Glashow[7]: "Such a theory is technically possible to construct, but it is both ugly and experimentally false." Thus, it is necessary to use a larger symmetry group, and consequently to include more gauge fields.

Clearly, the simplest alternative is to assume the existence of one more new gauge field component in order to complete the SU(2) weak isotopic-spin group. The combined weak and electromagnetic interactions then could be unified under a new gauge symmetry group given by the product SU(2) × U(1). However, this choice clearly raises some disturbing questions. First, the existence of a new gauge field component corresponding to τ_3 implies an entirely new class of weak interactions for both the neutrino and the electron. The reason for this is that the new W_μ^3 component of the weak gauge field would have the same coupling g as the $W_\mu^\pm$ components because they belong to the same set of generators for the SU(2) group. Since

[6] S. L. Glashow, *Nucl. Phys.* **22**, 579 (1961).
[7] S. L. Glashow, *Rev. Mod. Phys.* **52**, 539 (1980).

τ_3 has the diagonal matrix representation

$$\tau_3 = \begin{pmatrix} 1 & 0 \\ 0 & -1 \end{pmatrix} \quad , \qquad \text{(VIII - 12)}$$

the new gauge field will contribute a term to the kinetic energy. which has the form

$$\text{K.E.} = -ig\bar{L}\gamma^\mu W^3_\mu \tau_3 L \quad . \qquad \text{(VIII - 13)}$$

Thus, we see that in order to achieve unification, the Weinberg-Salam model predicts that there is a new weak interaction known as a "neutral current" interaction. Since the electron is a weak isotopic-spin state, the new interaction will contribute a weak force between electrons in addition to the familiar Coulomb force. This extra weak force will produce parity violating effects in ordinary electromagnetic phenomena, such as atomic spectra. The neutral current interaction will also allow a new type of elastic scattering for neutrinos, such as $\nu_\mu + e \rightarrow \nu_\mu + e$, which is forbidden in the old weak theory. At the same time, there was no experimental evidence for such neutral current interactions. In fact, the earlier gauge theories were constructed so as to explicitly exclude the possibility of neutral currents.

The existence of a new weak gauge field produces another very essential complication. In the old theory, the electric charge was measured by assuming that the total force between electrons was given by Coulomb's law. With the introduction of a new interaction, the old definition of the electric charge is no longer valid. The old electric charge and the old vector potential actually contain a contribution from the new weak interaction. Thus, the true electromagnetic potential A_μ^{em} is not just the gauge field A_μ of the U(1) group, but must be some linear combination of the U(1) gauge field and the new W^3_μ field of SU(2).

We can determine the correct form of A_μ^{em} by examining the purely neutral interaction terms in the kinetic energy. As we noted before, these terms would contribute to the elastic scattering of neutrinos. We first find from (VIII-10) and (VIII-13) that the term involving only the neutrino has the form

$$\bar{\nu}\gamma^{\mu}[qA_{\mu} + gW^3_{\mu}]\tfrac{1}{2}(1 + \gamma_5)\nu \quad . \tag{VIII - 14}$$

Since the neutrino has no electromagnetic interaction, the potential A^{em}_{μ} must be a linear combination of A_{μ} and W^3_{μ} that is "orthogonal" to (VIII-14). A simple way to calculate A^{em}_{μ} is to consider A_{μ} and W^3_{μ} to be orthogonal "unit vectors" in the linear vector space defined by the group generators themselves. Then it is straightforward to see that the combination

$$gA_{\mu} - qW^3_{\mu} \tag{VIII - 15}$$

is orthogonal to (VIII-14). In order to preserve the form of the Maxwell field tensor, we must also normalize (VIII-15). Thus the true electromagnetic potential is given by

$$A^{em}_{\mu} = (gA_{\mu} - qW^3_{\mu})/\sqrt{g^2 + q^2} \quad . \tag{VIII - 16}$$

The field combination in (VIII-14) also defines a new weak field designated by

$$Z^0_{\mu} = (qA_{\mu} + gW^3_{\mu})/\sqrt{g^2 + q^2} \quad . \tag{VIII - 17}$$

This field must be the "physical" neutral weak field instead of W^3_{μ} since it is clear from (VIII-16) that W^3_{μ} is not purely weak but also contains an electromagnetic contribution.

We can also easily calculate the new electric charge in terms of the coupling constant q and g. The physical charge is defined as the coupling of the electron to the A^{em}_{μ} field. We first solve for the A_{μ} and W^3_{μ} gauge fields in terms of the electromagnetic field A^{em}_{μ} and the weak Z^0_{μ} field and obtain

$$A_{\mu} = (qZ^0_{\mu} + gA^{em}_{\mu})/\sqrt{g^2 + q^2} \quad ,$$

$$W^3_{\mu} = (gZ^0_{\mu} - qA^{em}_{\mu})/\sqrt{g^2 + q^2} \quad . \tag{VIII - 18}$$

By using the expression for A_μ in the right-handed electron coupling in the kinetic energy (VIII-10), we obtain

$$\mathrm{i}\left(qg/\sqrt{g^2+q^2}\right)\bar{R}\gamma^\mu A_\mu^{\mathrm{em}} R + \mathrm{i}\left(q^2/\sqrt{g^2+q^2}\right)\bar{R}\gamma^\mu Z_\mu^0 R . \quad \text{(VIII - 19)}$$

We can easily read off the true electric charge as the coefficient of A_μ^{em},

$$e = qg/\sqrt{g^2+q^2} \quad . \quad \text{(VIII - 20)}$$

Since the coupling constant q associated with the U(1) gauge field is not the true electric charge, it is now called the weak "hypercharge".

The coupling of the Z_μ^0 field to the neutrino and the electron can be determined in the same way as the electric charge. Using (VIII-18) in the purely neutrino kinetic energy term (VIII-14), we find that the coefficient of the Z_μ^0 field is equal to

$$g_\nu^0 = \sqrt{g^2+q^2} \quad . \quad \text{(VIII - 21)}$$

The coupling of the Z_μ^0 field to the electron is somewhat more complicated because the left and right-handed couplings are different. From (VIII-19), we see that the right-handed coupling is equal to

$$g_{e_\mathrm{R}}^0 = q^2/\sqrt{g^2+q^2} \quad . \quad \text{(VIII - 22)}$$

To find the left-handed electron coupling, we extract the pure e_L terms from (VIII-10) and (VIII-13),

$$-\mathrm{i}q\bar{e}_\mathrm{L}\gamma^\mu A_\mu e_\mathrm{L} + \mathrm{i}g\bar{e}_\mathrm{L}\gamma^\mu W_\mu^3 e_\mathrm{L} \quad . \quad \text{(VIII - 23)}$$

Using (VIII-18) once again, we can read off the coefficient of the left-handed coupling to the Z_μ^0 field and obtain

$$g_{e_\mathrm{L}}^0 = (q^2-g^2)/\sqrt{g^2+q^2} \quad . \quad \text{(VIII - 24)}$$

The different couplings of the Z_μ^0 field for the left and right-handed spin polarization states of the electron are, of course, just a result of the parity violation in the weak interaction.

The weak and electromagnetic gauge fields are now completely unified. What is most interesting about the unification is the mixing of the U(1) and SU(2) gauge fields that was necessary to construct the physical electromagnetic potential. We began with a product of disconnected symmetry groups and ended up by unifying them through a mixing of gauge fields. The reason for the mixing, of course, has nothing to do with gauge theory *per se*. It was built in "by hand" through the identification of the leptons as the appropriate doublets and singlets of weak isotopic-spin.

8.4 Symmetry Breaking and Gauge Field Masses

To complete the building of the Weinberg-Salam model, we must break the gauge symmetry of the weak interaction and generate masses for the $W^{\pm}$ and Z^0 gauge fields. In chapter VII, we saw how the symmetry of the gauge fields can be broken through the interaction of the gauge fields with the self-coherent ground state of a Higgs field. Thus, we need to apply the same technique to the Weinberg-Salam model.

We noted previously that the gauge field mass term came directly from the ground state kinetic energy of the Higgs field. Thus we need only to calculate the kinetic energies of the $W^{\pm}$ and Z^0 fields using the "broken" form of the gauge potential

$$W'_\mu = A_\mu - \frac{1}{q}\partial_\mu \Phi \quad , \qquad \text{(VIII - 25)}$$

where Φ is again the coherent phase of the Higgs field. The new forms of the $W^{\pm}$ and Z^0 kinetic energies are given by

$$\left(g^2 |W_\mu^{\pm}|^2 + g_0^2 |Z_\mu^0|^2\right) |\phi_0|^2 \qquad \text{(VIII - 26)}$$

which we recognize as mass terms for the $W^{\pm}$ and Z^0 fields. We can easily read off the masses of the broken gauge fields,

$$M_W = g\sqrt{|\phi_0|^2} \quad ,$$

$$M_{Z^0} = g_0\sqrt{|\phi_0|^2} \quad , \qquad \text{(VIII - 27)}$$

where $|\phi_0|$ is the magnitude of the Higgs field.

In order to guarantee that the photon has zero mass, we assume *ab initio* that the Higgs ground state is electrically neutral. It must, however, carry weak isotopic-spin and weak hypercharge in order to interact with the $W^{\pm}$ and Z^0 fields. We note also that the symmetry breaking has provided the $W^{\pm}$ and Z^0 fields with a new component of spin. It is well known that massless spin-1 fields such as the electromagnetic field have only two spin components, namely parallel and antiparallel to the momentum which is a requirement of Lorentz invariance. A massive vector field must have an additional transverse spin component. The missing spin component arises from the Higgs phase term in the definition of the broken gauge field W'_{μ} in (VIII - 25). The phase term violates the so-called transversality requirement $\partial^{\mu} A_{\mu} = 0$ for zero mass fields. This result has motivated the picaresque saying among elementary particle physicists that the $W^{\pm}$ and Z^0 have gained weight and a transverse spin component as well by "eating the Higgs field".

8.5 The Weinberg Angle

It might appear from the preceding discussion that there are too many unknown quantities in the Weinberg-Salam theory to obtain any useful results. One would expect a truly unified theory to have only a single coupling constant rather than the two constants g and q for the SU(2) and U(1) gauge groups. The masses for the $W^{\pm}$ and Z^0 also involve the unknown Higgs ground state value $|\phi_0|$. Unfortunately, the quantity $|\phi_0|$ cannot be directly calculated. We noted earlier in chapter VII that the value $|\phi_0|$ depends explicitly on the parameters in the Higgs potential. Although the parameters can be found for a superconductor, their values cannot be determined in the Weinberg-Salam theory because not enough is known about the detailed properties of the Higgs field itself. However, all is not lost. It is still possible to make specific predictions for the $W^{\pm}$ and Z^0 masses.

The first step is to combine the two coupling constants g and q into a single new parameter called the Weinberg angle θ_W. The Weinberg angle is defined from the normalized forms of A^{em} and Z^0 in (VIII-16) and (VIII-17),

$$\sin \theta_W = q/\sqrt{g^2 + q^2} \quad ,$$

$$\cos \theta_W = g/\sqrt{g^2 + q^2} \quad , \tag{VIII - 28}$$

which shows that the mixing of the A_μ and W_μ^3 fields can be interpreted as a rotation through the Weinberg angle. We can next eliminate any need for the quantity $|\phi_0|$ by relating the new coupling constant g to the measured value of the old Fermi constant G from the V–A theory. The relation is given in (VIII-11), which shows why a large value of the W mass yields a very small Fermi coupling G even when the true coupling g is not small. The $W^\pm$ mass can now be expressed as

$$\begin{aligned} M_W^2 &= \frac{g^2}{2G} \\ &= \frac{e^2}{2G \sin^2 \theta_W} \\ &= \frac{(37.4 \text{ GeV})^2}{\sin^2 \theta_W} \quad . \end{aligned} \tag{VIII - 29}$$

The mass of the Z^0 is then related to the $W^\pm$ mass by

$$M_{Z^0}^2 = \frac{M_W^2}{\cos^2 \theta_W} \quad . \tag{VIII - 30}$$

The mass formulae for the $W^\pm$ and Z^0 still involve the unknown Weinberg angle but they do yield specific lower limits. We see that the $W^\pm$ must be heavier than 37.4 GeV and that the Z^0 mass is greater than that of the $W^\pm$. These large lower limits for the masses explain very obviously why no evidence for the W boson was seen in the early measurements of weak decays.

Is it possible to make better predictions for the masses? Obviously, the solution is to measure the value of the Weinberg angle. However, a measurement of θ_W also requires a direct measurement of neutral current interactions. This is clear from the definition

(VIII-28) of the Weinberg angle as the parameter which determines the mixing of the gauge fields. In order to obtain θ_W, the values and relative sign of the coupling constants g and q must be determined. However, g and q are not directly measurable. One must actually measure the interaction rates for both $W^{\pm}$ and Z^0 exchange processes and then extract q, g and θ_W. Thus, a measurement of the Weinberg angle is also a test of a basic prediction of the Weinberg-Salam theory, namely the existence of neutral current interactions.

8.6 Renormalization and Revival

It can be seen today with some hindsight that the unification of the weak and electromagnetic interactions in the Weinberg-Salam theory was precisely the remedy needed to cure the one remaining disease of the "non-renormalizable" infinities in gauge theory. However, it could not be proved at the time and the general response of the community of physicists to the Weinberg-Salam theory was best described some years later by Sidney Coleman[8]: "Rarely has so great an accomplishment been so widely ignored."

The "Dark Age" of gauge theory persisted for another four years until 1971 when the final breakthrough occurred. A brilliant young graduate student, Gerard 't Hooft[9] at the University of Utrecht in the Netherlands, invented a new technique for proving that spontaneously broken gauge theories are renormalizable. The technique developed by 't Hooft and later refined by others[10] is extremely subtle and complex. However, the effect of the cure for the "non-renormalizable" infinities can be briefly stated, at least qualitatively. The unification of electromagnetism and the weak interaction in the Weinberg-Salam theory introduces a Z^0 gauge field whose coupling constant is related in a very special way to the couplings of the charged weak and electromagnetic interactions. In the scattering of a neutrino and an

[8] S. Coleman, *Science* **206**, 1290 (1979, N. Y.).
[9] G. 't Hooft, *Nucl. Phys.* **B35**, 167 (1971).
[10] B. W. Lee, and J. Zinn-Justin, *Phys. Rev.* **D5**, 3121 (1972).

antineutrino

$$\nu + \bar{\nu} \longrightarrow W^+ + W^-$$

which we mentioned in chapter VI, the exchange of a Z^0 provides a new contribution shown in Fig. (8.1). It can be shown that this new term provides just the right cancellation effect so that the cross-section no longer increases indefinitely with energy. The diagram in Fig. (8.1) also shows that the non-Abelian nature of the Yang-Mills gauge field is essential. In an Abelian gauge theory like electromagnetism, the $W^\pm$ and Z^0 fields would not interact directly with each other. For other previously "infinite" reactions, such as

$$e^+ + e^- \longrightarrow W^+ + W^- \quad ,$$

the contribution of the electromagnetic field is also necessary to eliminate the infinities. Thus, it was the intimate connection between unification and renormalization that provided the final piece of the gauge puzzle.

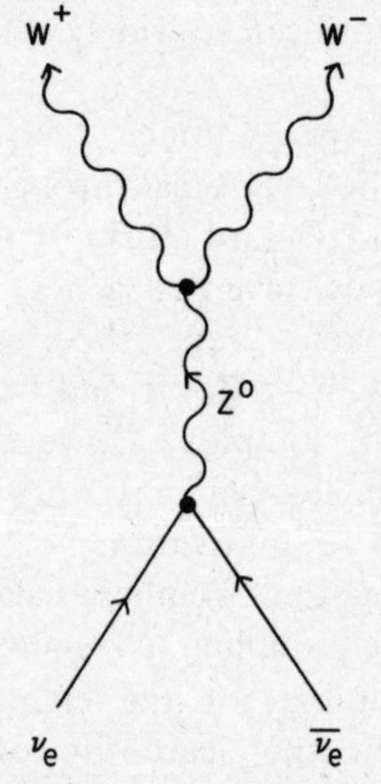

Fig. 8.1 New contribution to the scattering of a neutrino and an antinuetrino from Z^0 exchange in the Weinberg-Salam unified theory. This term generates new higher-order terms which cancel the "non-renormalizable" infinities.

The Weinberg-Salam theory was extended to include quarks as well as leptons by Glashow, Iliopoulis and Maiani[11]. They predicted the existence of a new quark flavor called "charm". The charmed quark was the intended partner of the strange quark in an isotopic-spin $\frac{1}{2}$ doublet[a] like the u and d quarks,

$$\begin{pmatrix} u \\ d' \end{pmatrix}, \begin{pmatrix} c \\ s' \end{pmatrix} .$$

Given that the quarks and leptons both occur in isotopic-spin doublets as predicted, their weak interactions are essentially the same. The lepton current (VIII-5) can be converted into a quark current by simply re-labelling the spinors. However, since the quarks masses are not zero like the neutrino, the quark doublets involve both right and left-handed states.

The revival of the Weinberg-Salam theory was the beginning of modern gauge theory. The successful` experimental tests soon followed. A new family of particles carrying the charm quantum number was discovered in e^+e^- collisions in the November of 1974 at the Stanford Linear Accelerator Center, an event which is now hailed by elementary particle physicists as the "November revolution"[12]. In 1973, the prediction of neutral current interactions was first confirmed by the observation of $\nu_\mu e^-$scattering in the Gargamelle bubble chamber at CERN[13]. The measurement of parity violation in high energy scattering of electrons was performed in 1979 in a beautiful experiment which utilized the 2-mile long SLAC linear accelerator as part of the experimental apparatus[14]. Some very difficult measurements have also been made to detect the tiny parity-violating optical effects in excited atoms[15]. These results and others tests of the Weinberg-Salam

[a]The d' and s' states are mixtures of the pure d and s quark states.

[11]S. L. Glashow, J. Iliopoulis and L. Maiani, *Phys. Rev.* **D2**, 1285 (1970).
[12]W. K. H. Panofsky, *Contem. Phys.* **23**, 23 (1982).
[13]G. Myatt, *Proc. Intl. Symp. on Electron and Photon Interactions at High Energies,* 1973 ed. H. Rollnik and W. Pfeil. p. 389(North-Holland, Amsterdam).
[14]C. Y. Prescott *et al., Phys. Lett.* **84B**, 524 (1979).
[15]E. D. Comminis *et al., Phys. Rev.* **D24**, 1134 (1981).

theory have been summarized in recent review articles[16]. On the basis of these successful experimental tests of the unified theory, the Nobel prize in physics was awarded to Glashow, Weinberg and Salam in 1979.

Clearly a crucial test of the Weinberg-Salam theory is the direct observation of the $W^{\pm}$ and Z^0 gauge bosons. Measurements of the charged and neutral weak coupling constants in many different weak interactions have been made. Thus it is possible to predict the mass values of the $W^{\pm}$ and Z^0 with reasonable accuracy. The results from many experiments have all yielded a consistent value for the Weinberg angle of $\sin^2\theta_W = 0.23$. By using this value in the equations for the $W^{\pm}$ and Z^0 masses, one obtains predictions[b] of approximately 77 GeV and 88 GeV respectively, which can be compared to the mass of an atom of gold!

It is clear that the $W^{\pm}$ and Z^0 particles have never been seen before in high-energy collisions because of their high masses. Given enough energy, would they appear like ordinary particles? We discussed earlier in chapter V that gauge theories with non-Abelian groups like SU(2) have no general plane-wave solutions because of the non-linear terms in the wave equation. Does this mean that the $W^{\pm}$ and Z^0 will not behave like particles or, even worse, not be observable? We can find part of the answer by estimating how important the non-linear terms are in the wave equation. We first note that the Yang-Mills wave equation can be applied to the $W^{\pm}$ or Z^0 fields by simply adding the appropriate mass term. The general form of the wave equation remains the same in the Weinberg-Salam theory because the Lagrangian terms such as $F_{\mu\nu}F^{\mu\nu}$ are unchanged by the gauge symmetry breaking. We see from the wave equation that the coupling constant g multiplies the non-linear terms.

[b]The calculation of the $W^{\pm}$ and Z^0 masses also require higher-order corrections which increase the predicted values to 79 GeV and 91 GeV respectively [Green and Veltman, 1980].

[16]G. Myatt, *Rep. Prog. Phys.* **45**, 1 (1982).

The value of g can be obtained from

$$g = \frac{e}{\sin\theta_W} \approx 2e \quad , \qquad \text{(VIII - 31)}$$

which shows that g is of the same order of magnitude as the electromagnetic coupling. Thus, the contribution from the non-linear terms is very small and we would expect the $W^{\pm}$ and Z^0 to behave like ordinary particles. The small value of g also indicates why the perturbative techniques used so successfully in QED can be used in calculations for the Weinberg-Salam theory.

8.7 The Electron Mass

The Weinberg-Salam model also provides some insight into how the electron mass may be generated. Due to the schizophrenic classification of the electron as both a weak isotopic-spin doublet and singlet, it is impossible to have an electron mass term in the Lagrangian which is invariant under SU(2). Thus, one can apply Occam's razor and see if the electron mass can also be generated by interaction with the Higgs field ground state.

There is no general gauge theory principle like minimal coupling that tells us how the leptons interact with the scalar field. It must be put in by hand. If it is assumed that the Higgs field is also a weak isotopic-spin doublet, the simplest SU(2)-invariant interaction has the form[17]

$$\mathcal{L}_e = G_e(\bar{R}\bar{\phi}_0 L + \bar{L}\phi_0 R) \quad , \qquad \text{(VIII - 32)}$$

where G_e is a new unknown coupling constant. There are two constraints which must be satisfied for this interaction. First of all, this new interaction of the Higgs field must not lead to a non-zero mass for the photon. At the same time, the interaction must also maintain a zero mass for the neutrino. Both of these requirements can be satisified if the Higgs ground state is equivalent to a weak

[17] C. Quigg, *Introduction to Gauge Theories of the Strong, Weak and Electromagnetic Interactions* (Fermilab-Conf-80/64-THY, July 1980).

isotopic spinor of the form

$$\phi_0 \sim \begin{pmatrix} 0 \\ 1 \end{pmatrix} \quad . \tag{VIII-33}$$

We note that choosing this particular representation for the isotopic-spin content of ϕ_0 does not affect any of our previous calculations because they were independent of the detailed form of ϕ_0. Thus, the interaction (VIII-32) becomes

$$\mathcal{L}_e = G_e|\phi_0|\bar{e}e \tag{VIII-34}$$

so that the mass of the electron is simply

$$m_e = G_e|\phi_0| \quad . \tag{VIII-35}$$

The value of the mass is not predicted, of course, since the coupling G_e is unknown. However, we see that the gauge symmetry breaking mechanism is capable, in principle, of generating the masses of both the gauge fields and the leptons.

8.8 Discussion

Although we have presented only a brief discussion of the Weinberg-Salam unified gauge theory, it should be evident that the theory represents the culmination of many years of difficult experimental and theoretical research. In chapter VI, we barely touched on the brilliant analysis and the many years of hard work required to untangle the experimental details of weak decays just to be able to see the underlying gauge theory structure. It should be remembered that when Weinberg and Salam proposed their theory, there was almost no experimental data available on weak interactions at high energies. There were also no high energy neutrino beams and no direct evidence for neutral current interactions. It is therefore remarkable that some of the most successful predictions of the theory involve results from high energy neutrino scattering experiments. The sequence of developments leading up the Weinberg-Salam unified theory has been aptly described by Coleman[8] :

There is a popular model of a breakthrough in theoretical physics. A field of physics is afflicted with a serious contradiction. Many attempts are made to resolve the contradiction; finally, one succeeds. The solution involves deep insights and concepts previously thought to have little or nothing to do with the problem. It unifies old phenomena and predicts unexpected (but eventually observed) new ones. Finally, it generates new physics: the methods used are sucessfully extended beyond their original domain.

The Weinberg-Salam theory answers many of the most fundamental questions of physics, but at the same time it raises even more questions. First, does the theory truly unify the weak and electromagnetic interactions in the sense that there are no quantities which have to be determined from outside the theory? Clearly the value of the Weinberg angle cannot be calculated from the theory. Thus the amount of the SU(2) and U(1) mixing is not *a priori* predicted by the theory. It was also evident that the unification depended crucially on how the fundamental leptons are classified as weak isotopic-spin doublets and singlets. It was only then that the mixing of the gauge fields and the proper form of the physical fields and charges could be determined. Thus we see that the unification achieved by the Weinberg-Salam theory is a unification of the structure of the gauge fields but not of the particles which are the sources of the fields.

The most important feature of the Weinberg-Salam theory may ultimately be the symmetry breaking Higgs field. The Weinberg-Salam theory imposes very few constraints on the Higgs field so that it is difficult to predict many of its properties[18]. We saw earlier that the Ginzburg-Landau form of the Higgs potential applies equally well to elementary particles and condensed matter physics. Yet a detailed understanding of the dynamics of the Higgs field may be even more important for gauge theory than the gauge fields themselves. The Higgs field is intimately involved in two of the most revolutionary aspects of the Weinberg-Salam theory. It shows for the first time how mass is generated through an interaction

[18] J. Ellis, M. K. Gaillard and D. V. Nanopoulos, *Nucl. Phys.* **B106**, 292 (1976).

and it plays the role of a new type of vacuum in gauge theory. The latter property has raised questions about the origin of the Higgs field. In the Weinberg-Salam theory, the Higgs field is analogous to an old-fashioned "aether" which pervades all space-time. It acts like a continuous background medium even at very short distances. Clearly we would like to know where this aether came from. We saw in the case of the superconductor that the Higgs field was a composite system of electrons bound into Cooper pairs. The macroscopic coherent phase of the Higgs field was the result of the unusual binding force between the electrons. Could the Higgs field for the Weinberg-Salam theory also be a composite system of bound particles? Unfortunately, the analogy with the superconductor breaks down because there is no background atomic lattice in the Weinberg-Salam theory to provide the binding force. Thus, if the Higgs field is a composite system, the binding would have to be the result of a new force which has a much shorter range than the weak interaction. This hypothesis has motivated the investigation of new hypothetical gauge theories of superstrong forces.

CHAPTER IX

COLOR GAUGE THEORY

The trouble with nuclear physics is that it is not a subject that develops logically, but a collection of topics. It is not possible to start axiomatically and proceed from there.

L. R. B. Elton, 1958[1]

9.1 Introduction

One of the most promising developments of modern gauge symmetry is the emergence of a new theory which has become the leading candidate for a fundamental theory of the strong nuclear force. The theory is based on a new hypothetical quantum number called "color" which is carried by the quarks. Thus the new gauge theory of the strong interaction is called "color gauge theory" or "chromodynamics". Unlike the early theories involving meson exchange which we mentioned in chapter II, chromodynamics is postulated to be an exact local gauge theory. It is a direct descendent of the original Yang-Mills theory of isotopic-spin symmetry except that it has been applied to a new sub-nuclear domain of colored quarks and gauge fields.

We saw previously that the Yang-Mills theory failed to describe the strong interaction because it could not explain the short range of the nuclear force. We also found that the gauge potential fields in any unbroken gauge theory necessarily have zero mass. Thus why do we now believe that the strong interaction is a gauge theory with a new quantum number called color? The arguments for a color gauge theory are based on two important developments in elementary particle physics. The first is the great success of the quark model in

[1] L. R. Elton, *Introductory Nuclear Theory* (Interscience, New York, 1958).

describing the huge number of hadrons as bound states of a small number of elementary quarks. This showed that quarks, rather than mesons or nucleons, are the true sources of the strong nuclear force. However, no one has yet observed the production of free quarks in high energy collisions even though their masses are believed to be relatively small. This elusive nature of the quarks is related to the second piece of evidence for a non-Abelian gauge theory. An entirely new property of gauge theory has been uncovered which is called "asymptotic freedom". It is a feature which is unique to non-Abelian gauge theories and it shows how quarks may be tightly bound inside mesons and nucleons with a force which appears to be short-ranged but does not break gauge symmetry.

In this chapter, we will present a simple introduction to the basic physical concepts in chromodynamics. It is important to keep in mind that neither the theoretical predictions nor the experimental tests of chromodynamics have yet achieved the level of either QED or the Weinberg-Salam theory. The reason for this is that any gauge theory of the strong interaction must deal with the non-linear terms in the Yang-Mills equations. Thus in our treatment, we will rely primarily on intuitive physical arguments to understand the salient features of chromodynamics. In the following discussion, we will refer to chromodynamics by the familiar quantum acronym "QCD" even though we are using a semiclassical description.

9.2 Colored Quarks and Gluons

The new internal quantum numbers in QCD are three colors[2,3] which are conventionally called red, green and blue. These color charges are assumed to be carried by three quark states which form a fundamental basis for the QCD gauge group SU(3) in the same way that the proton and neutron form a basis state of the isotopic-spin group SU(2). Why do we need three fundamental colors instead of just two states like isotopic spin? Originally, color was invented as a purely *ad hoc* device to solve a problem in the quark model.

[2] M. Y. Han and Y. Nambu, *Phys. Rev.* **139**, 1006 (1965).
[3] O. W. Greenberg and C. A. Nelson, *Phys. Rep.* **32C**, 69 (1977).

Quarks are considered to obey Fermi-Dirac statistics like electrons. The Pauli principle forbids the existence of states with three identical quarks, such as uuu, which are completely symmetric in both spin and space degrees of freedom. However, such a state actually exists, the Δ^{++} resonance. To resolve this contradiction, it was postulated that each quark carried a new quantum number, color, which came in a minimum of three different shades. This original definition of color was global like the old isotopic spin.

Some of the mathematical tools necessary for the SU(3) color group can be borrowed directly from the well-known SU(3) classification of the hadrons in the old quark model[4]. The three colors are assigned to the quark triplet representation as shown in Fig. (9.1). There are two different fundamental bases for SU(3), a triplet and a complex conjugate antitriplet; thus, the anticolors (cyan, magenta and yellow) are carried by three antiquarks which are assigned to the antitriplet. With the choice in Fig. (9.1), the red and green quarks form a "color isotopic-spin" doublet and the blue quark corresponds to a "color hypercharge" state. The exact placement of the three colors in Fig. (9.1) is completely arbitrary.

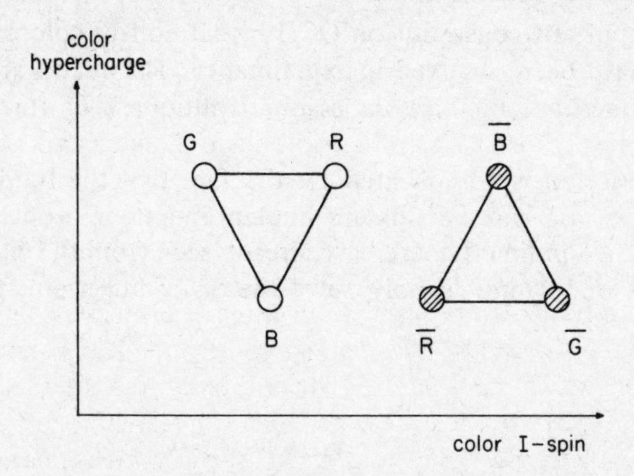

Fig. 9.1 The three quark colors, red, green and blue, belong to a fundamental triplet of SU(3). The anticolors belong to an antitriplet. The colors red and green form a color isotopic-spin pair and the color blue corresponds to a color hypercharge.

[4] M. Gell-Mann and Y. Ne'eman, *The Eightfold Way* (Benjamin, New York, 1964).

The gauge potential fields in QCD are called colored "gluons" because they provide the binding force between the quarks. Since there are eight SU(3) generators, it is conventional to say that there are eight different gluon fields. The gluon fields act like isotopic-spin raising and lowering operators except that they change the color of the quarks. For example, let us consider a red quark which is somehow accelerated so that it emits a gluon and then turns green as shown in Fig. (9.2). This process is analogous to bremsstrahlung in electrodynamics. The emitted gluon changes the quark color from red to green. A useful mnemonic is to visualize the gluons as carrying both a color and an anticolor simultaneously. Thus in the above example, the gluon can be considered as carrying red and anti-green colors. During the emission, the gluon picks up the red color from the initial quark state and leaves behind the green color in the final quark state. This two-color picture of gluons is extremely convenient but it should be remembered that it is motivated by the fact that color gauge fields are actually operators in the internal color space.

9.3 Colorless Quark Systems

An important constraint on QCD is that no free colored quarks or gluons have been observed in experiments. The quarks are bound inside the mesons and baryons as quark-antiquark or three quark states with net color charge of zero. Thus our understanding of the strong interaction is complicated by the fact that the fundamental color charges and gauge fields are hidden and their properties can only be studied indirectly. We have already seen from the Yang-Mills theory that one cannot simply solve the Schrödinger equation and

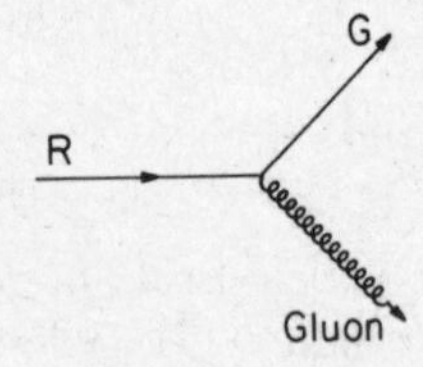

Fig. 9.2 A red quark is accelerated and changes color by emitting a gluon through color "bremsstrahlung". The gluon carries off the red and antigreen colors.

find general analytic solutions for these two cases even in the semi-classical formalism. However, we can learn something interesting by looking at some simplified solutions. Let us see what assumptions are necessary for colorless Coulomb-like solutions to exist. First of all, it is evident that any solution for which the color vectors of all quarks point in the same direction are excluded because the color vectors obviously would not cancel each other and give a net color charge of zero. However, the SU(3) group has two generators, F_8 and F_3, which commute with each other. The eigenvalues of these generators correspond to color "hypercharge" and the third component of color "isotopic spin". If we take the special case where the color charges are only linear combinations of F_8 and F_3, they will automatically commute. The color charges then can be added like ordinary numbers and we need only find the appropriate combination for net color charge zero.

Let us consider the three-quark system first. As shown in Fig. (9.1), we will identify the color blue with hypercharge, and red and green with the two components of color isotopic spin. The color charges can be written in terms of F_8 and F_3 as follows[5]

$$q_B = -\bar{\psi} \frac{1}{\sqrt{3}} F_8 \psi \quad , \tag{IX - 1a}$$

$$q_R = \bar{\psi} \frac{1}{2\sqrt{3}} \left(F_8 + \sqrt{3} F_3 \right) \psi \quad , \tag{IX - 1b}$$

$$q_G = \bar{\psi} \frac{1}{2\sqrt{3}} \left(F_8 - \sqrt{3} F_3 \right) \psi \quad . \tag{IX - 1c}$$

The sum of the color charges, red-blue-green, is zero as required for the ground state of a baryon. Any other combination of three charges, such as red-red-blue, clearly gives a non-zero total color charge. The choices, (IX-1), are not unique since they were determined by the arbitrary assignment of blue to correspond to hypercharge. We could have chosen any combinations which leave the

[5]P. Sikivie and N. Weiss, *Phys. Rev.* D18, 3809 (1978).

total color charge unchanged. Using the above choice of color charges, the colorless Coulomb potential for three-quarks can be written[5]

$$A_i = 0 \qquad (i = 1, 2, 3) \quad ,$$

$$A_0 = \frac{1}{4\pi\sqrt{3}} \left\{ F_8 \left(\frac{-1}{|\mathbf{x} - \mathbf{x}_B|} + \frac{1}{2|\mathbf{x} - \mathbf{x}_R|} + \frac{1}{2|\mathbf{x} - \mathbf{x}_G|} \right) \right.$$

$$\left. + F_3 \frac{\sqrt{3}}{2} \left(\frac{1}{|\mathbf{x} - \mathbf{x}_R|} - \frac{1}{|\mathbf{x} - \mathbf{x}_G|} \right) \right\} \quad , \qquad \text{(IX - 2)}$$

where $\mathbf{x}_R$, $\mathbf{x}_B$ and $\mathbf{x}_G$ are the positions of the colored quarks.

The quark-antiquark system is much simpler. The anticolor charges are obtained from (IX-1) by changing the signs. The only colorless quark-antiquark systems consist of a color charge and its exact anticolor charge, such as red-antired. For convenience, we can choose the color charge to point along F_8 and the anticolor in the opposite direction. Any other color direction is equivalent since a gauge transformation would rotate both the color and anticolor in the same way. The colorless quark-antiquark potential can be written[5]

$$A_i = 0 \qquad (i = 1, 2, 3) \quad ,$$

$$A_0 = \frac{1}{4\pi\sqrt{3}} F_8 \left\{ \frac{1}{|\mathbf{x} - \mathbf{x}_1|} - \frac{1}{|\mathbf{x} - \mathbf{x}_2|} \right\} \quad , \qquad \text{(IX - 3)}$$

where $\mathbf{x}_1$ and $\mathbf{x}_2$ are the positions of the quark and antiquark.

The solutions (IX-2) and (IX-3) show that it is relatively straightforward to find exact static potentials for colorless quark-antiquark and three-quark systems which are linear combinations of Coulomb potentials just as in QED. Do these Coulomb solutions have any relevance for real mesons and baryons? We might predict that the quark-antiquark bound state has energy levels analogous to those of positronium. This rather naive prediction is obviously not true for the majority of mesons with masses less than about 2 GeV. It does appear to hold at least qualitatively for the bound states of the charmed quark[6,7]. The energy levels of the excited states of "charmonium" resemble a heavy form of positronium, although the precise values of the energies do not agree exactly with a simple Coulomb potential[8]. This serendipitous agreement suggests that the Coulomb potential provides a reasonable approximation for the color binding force under certain conditions, namely, that the charmed quark is sufficiently massive (about 1.5–2 GeV) so that the motion of the quarks is nonrelativistic and the quark and antiquark are close together. The deficiency in this explanation is that it does not give us any insight into the most important question in the quark model: if hadrons are bound states of quarks, why are free colored quarks and gluons not observed in the breakup or decay of hadrons? We know from QED that charged particles bound in a Coulomb potential can always be liberated with relatively little energy. In constrast, high energy collisions between hadrons at energies hundreds of times larger than their masses have failed to produce any free quarks or gluons. This situation might be explained by postulating unprecedented and enormous binding energies for quarks, but that would obviously conflict with the apparently simple energy spectrum of charmonium. The solution to this mystery must be associated with an entirely new property of QCD.

9.4 Asymptotic Freedom

The existence of a new feature of the strong interaction was originally indicated by a series of beautiful experiments[9] which are the high energy counterparts of the classic Rutherford experiments with alpha particles. High energy electrons from the 2-mile long Stanford linear accelerator were scattered on proton targets in order to study the structure of the proton. The scattering process, which is known as

[6] T. Appelquist, R. M. Barnett and K. Lane, *Ann. Rev. Nucl. Sc.* **28**, 287 (1978).
[7] R. F. Schwitters, Sc. Am. **237**, 56 (1977).
[8] B. Sheehy and H. C. von Baeyer, *Am. Jour. Phys.* **49**, 429 (1981).
[9] H. Kendall, *Proc. Vth Intl. Symp. Electron and Photon Interactions at High Energies* (Cornell University, 1971).

"deep-inelastic" scattering is shown in Fig. (9.3). The interaction is mediated by the exchange of a virtual photon which probes the internal structure of the target proton. Through the uncertainty principle, the momentum Q of the virtual photon can be related inversely to the size of the region which is seen by the probe. For Q values larger than a few GeV/c, the size is smaller than the radius of the proton, hence the name "deep-inelastic". Like the classic Rutherford experiment, the deep-inelastic scattering results showed that the proton and neutron consist of point-like charged constituents, which were later identified with quarks. In addition, it was observed that as Q was increased and the size of the volume being probed decreased, the effective strength of the interaction between the quarks appeared to become weaker. The quarks acted as if they were essentially free particles in the proton. This phenomenon was originally postulated to exist by J. Bjorken[10] and it was thus called "Bjorken scaling". This behaviour contradicted the traditional idea of the strong force becoming even stronger at short distances. Bjorken scaling seemed to imply that perhaps as one probed at ever smaller distances with even higher energies, that the quarks would ultimately behave like free particles. This revolutionary idea became known as "asymptotic freedom".

In order to understand the origin of asymptotic freedom in a color gauge theory, we will first compare it with a much more

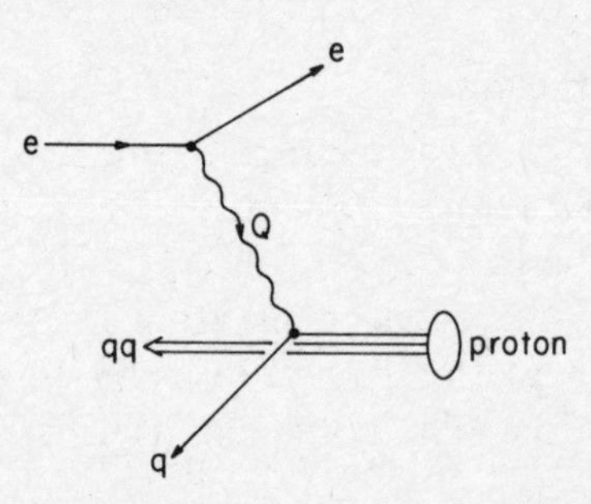

Fig. 9.3 Deep-inelastic scattering of high-energy electron on a proton. The interaction is mediated by the exchange of a virtual photon of momentum Q which probes the quarks in the proton.

[10] J. D. Bjorken, *Phys. Rev.* **179**, 1547 (1969).

familiar property of QED.[10,11,12,13,14] In QED, an electron is surrounded by a cloud of virtual electron-positron pairs which are produced from the vacuum by the electric field. These virtual pairs are the cause of the well-known Lamb shift in the energy spectrum of the hydrogen atom. The virtual positrons are attracted slightly closer to the electron while the virtual electrons are repelled. Thus the space around the electron is polarized like a dielectric medium as illustrated in Fig. (9.4). This polarization results in a partial shielding of the "bare" electron charge by the virtual positrons so that at large distances the effective charge of the electron appears to be smaller.

In QCD it is reasonable to expect a similar vacuum polarization from pairs of virtual quarks and antiquarks. Let us consider the effect of this polarization on the color force between two quarks in a proton, The magnitude of the force is determined by the effective color charge seen by each quark. If we decrease the distance between the quarks, each quark will penetrate further inside the

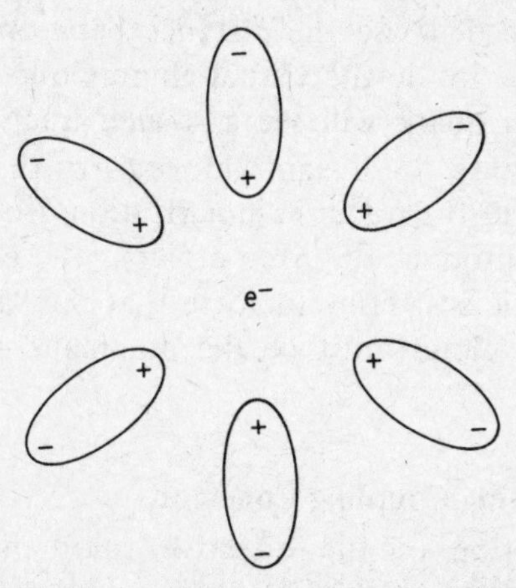

Fig. 9.4 Vacuum polarization of space around an electron by production of virtual electron positron pairs. The virtual electrons are repelled and the virtual positrons are attracted closer.

[11] Y. Nambu, *Sci. Am.* **235**, 48 (1976).
[12] E. Ma, *Am. Jour. Phys.* **47**, 873 (1979).
[13] W. Marciano and H. Pagels, *Nature* **279**, 479 (1979).
[14] G. 't Hooft, *Sci. Am.* **242**, 104 (1980).

cloud of virtual quark pairs surrounding the other quark and will see a larger effective color charge. Since the strength of the interaction depends on the net color charge seen by the quarks, the force between the quarks will become stronger. In deep-inelastic scattering, the high-Q interactions correspond to measurements of quark configurations where the separations are small. Thus, the strength of the interaction will appear to be stronger for large Q values in contradition to the experimental observation of asymptotic freedom. The reason for this disagreement is that we have not included the effects of the gluons. Unlike photons, the gluons carry color charge and therefore can interact directly with each other. This means that the color electric field around a quark can produce virtual gluons as well as quarks. Unlike the virtual quark-antiquark pairs, which have no net color, the virtual gluons carry part of the total color charge. We cannot calculate the exact distribution of the virtual gluon cloud, but we can reasonably assume that the virtual gluons spread the total color charge throughout some finite volume like the old-fashioned Thomson model of the atom. Now if we decrease the distance between the quarks again, each quark will be inside the virtual gluon cloud of the other quark. In this case, each quark will see a smaller fraction of the effective color charge because the distant gluons carry the remaining charge. Thus we see that the vacuum polarization from quark-antiquark pairs and gluons produce opposite effects. The experimental results from deep-inelastic scattering indicate that the "antiscreening" from the virtual gluon cloud must be the dominant effect at very small distances.

9.5 The Running Coupling Constant

The calculation of the effective color charge from vacuum polarization in QCD was originally done with perturbative quantum field theory techniques which are generalizations of those used in QED[15,16]. However, it has been pointed out[17,18] that the QCD

[15] D. Gross and F. Wilczek, *Phys. Rev. Lett.* **31**, 1343 (1973).
[16] H. D. Politzer, *Phys. Rev. Lett.* **30**, 13 (1973).
[17] T. D. Lee, *Is the Physical Vacuum a Medium?* (Colombia University, CU-TP-170, 1979).
[18] V. Weisskopf, *Physics Today* **34**, 69 (1981).

vacuum can also be treated semiclassically like a dielectric medium. In this picture, the potential between two charges is given by

$$V(r) = \frac{q_1 q_2}{4\pi\epsilon(r)r} \quad , \qquad \text{(IX - 4)}$$

where all of the vacuum polarization effects are absorbed into an effective dielectric constant $\epsilon(r)$ for the vacuum. The QCD dielectric constant has been calculated by Nielsen and Olesen[19,20] who used semiclassical techniques and obtained the same result as the quantum field theory calculation. Many of the steps used in the calculation are too lengthy to reproduce here so we will only summarize the physical arguments.

In order to obtain an effective dielectric constant for the QCD vacuum, we can follow the same approach used for a dielectric medium in electrodynamics. However, instead of applying an electric field in the vacuum, it is much simpler to use a magnetic field. The classical energy density of the medium in an external magnetic field is given by

$$E_0 = -\frac{1}{2} 4\pi\chi H^2 \quad , \qquad \text{(IX - 5)}$$

where χ is the magnetic susceptibility of the medium. The relation between χ and the permeability μ is given by

$$\mu = 1 + 4\pi\chi \quad . \qquad \text{(IX - 6)}$$

The dielectric constant ϵ is then obtained from the relation

$$\epsilon\mu = 1 \quad , \qquad \text{(IX - 7)}$$

where the velocity of light has been set to 1. This relation, which is not true for an ordinary polarizable medium, is essential to guarantee that the QCD vacuum is relativistically invariant. Thus we see that the dielectric constant $\epsilon(r)$ can be obtained from the energy of the vacuum by applying an external magnetic field.

[19] N. K. Nielsen and P. Olesen, *Nucl. Phys.* **B144**, 376 (1978).
[20] N. K. Nielsen, *Am. Jour. Phys.* **49**, 1171 (1982).

The QCD vacuum consists of virtual quarks, antiquarks and gluons. In a semiclassical picture, these are represented by the ground states of the spin-$\frac{1}{2}$ Dirac field and spin-1 zero-mass boson field. In QED, the ground state energy of the electromagnetic field is just equal to the zero-point energy of the quantized harmonic oscillator[21]. On the other hand, the ground state energy of the Dirac field is twice as large and has the opposite sign. The reason for this is that the commutation relations for the electromagnetic field operators must be replaced by anticommutators for the Dirac field which give the opposite sign. In addition, the Dirac field has four spin states while a massless spin-1 field has only two. It can be shown that the calculation for the energies of the Dirac or spin-1 ground states in an external magnetic field lead to eigenvalue equations which resemble harmonic oscillator problems. By solving these equations and summing over the oscillator modes, one then obtains the vacuum energy. Unfortunately, the summation turns out to be a badly divergent function of the mode energy and one therefore has to introduce cutoffs on the allowed energies which contribute to the final sum. Following this general procedure, Nielsen and Oleson obtained the vacuum energy equation

$$E_0 = -(-1)^{2s} \sum_{S_3} \left(\frac{S_3^2}{2} - \frac{1}{24} \right) \frac{Ve^2 H^2}{4\pi^2} \log\left(\frac{E_2}{E_1} \right) \qquad \text{(IX - 8)}$$

where H is the magnetic field strength, V is a volume used to define the density of oscillator states and S_3 is the spin component of the Dirac or spin-1 field along the magnetic field direction. The factor $(-1)^{2s}$ in front gives the correct sign for the Dirac or spin-1 vacuum energy. The quantities E_2 and E_1 are the upper and lower cutoffs on the oscillator modes energies which are necessary to keep the logarithm from giving an infinite value. For the external field H, the value of E_1 is determined by $|eH|$ which is the fundamental oscillator frequency. This cutoff eliminates modes with wavelengths greater than the fundamental oscillator wavelength.

The QCD vacuum energies for virtual quarks and gluons are

[21] L. I. Schiff, *Quantum Mechanics* (McGraw-Hill, New York, 1955).

now obtained from (IX - 8) by summing over the spin states and replacing the charge e with the appropriate SU(3) coupling. Let us first calculate the contribution from the virtual gluons. Summing over the two spin states ±1 gives

$$(E_0)_{\text{Gluon}} = -\frac{11Ve^2H^2}{48\pi^2}\log\left(\frac{E_2}{E_1}\right) \quad . \tag{IX - 9}$$

To determine the coupling of the virtual gluons to an external color magnetic field, we can use the Yang-Mills current J_ν derived in chapter V,

$$J_\nu = j_\nu - \text{i}g[A_\mu, F_{\mu\nu}] \quad , \tag{IX - 10}$$

where the commutator term is the pure gluon contribution. The interaction of this gluon current term with an external field $\widetilde{A}_\nu$ is given by

$$(J_\nu\widetilde{A}_\nu)_{\text{Gluon}} = \text{i}g\widetilde{A}_\nu[A_\mu, F_{\mu\nu}] \quad . \tag{IX - 11}$$

By writing the fields $\widetilde{A}_\nu$, A_μ and $F_{\mu\nu}$ as linear combinations of the SU(3) generators F_k, we obtain

$$\begin{aligned} g\widetilde{A}_\nu[A_\mu, F_{\mu\nu}] &= g(\widetilde{A}^i_\nu F_i)(A^j_\mu F^k_{\mu\nu}[F_j, F_k]) \\ &= gc_{jkl}\widetilde{A}^i_\nu A^j_\mu F^k_{\mu\nu}(F_iF_l) \quad , \end{aligned} \tag{IX - 12}$$

which shows that the gluon coupling is

$$gc_{jkl} \quad . \tag{IX - 13}$$

A simple way to evaluate this coupling is to choose the external color magnetic field to point along F_8 in the internal color space. The only SU(3) structure constants which contribute to the coupling are then[4] (also see the Appendix)

$$c_{458} = c_{678} = \frac{\sqrt{3}}{2} \quad . \tag{IX - 14}$$

Using these two coupling terms, we replace the charge e in (IX-9) with $g/\sqrt{3}$ and obtain for the gluon vacuum energy,

$$(E_0)_{\text{Gluon}} = -\frac{33 V g^2 H^2}{96\pi^2} \log\left(\frac{E_2}{E_1}\right) \quad . \qquad \text{(IX - 15)}$$

To calculate the vacuum energy from the virtual quarks, we sum (IX-8) over the four Dirac spin states. For the quark couplings to the external color magnetic field, we can use the color quark charges in (IX-1) which give quark couplings of $g/\sqrt{3}$, $g/2\sqrt{3}$ and $g/2\sqrt{3}$ along the F_8 direction. Summing over all these contributions, the vacuum energy is

$$(E_0)_{\text{Quark}} = N_f \frac{V g^2 H^2}{48\pi^2} \log\left(\frac{E_2}{E_1}\right) \quad , \qquad \text{(IX - 16)}$$

where N_f is the number of different types or flavors of quarks. The combined QCD vacuum energy for the virtual quarks and gluons is then

$$E_0 = -\frac{(33 - 2N_f)}{96\pi^2} V g^2 H^2 \log\left(\frac{E_2}{E_1}\right) \quad . \qquad \text{(IX - 17)}$$

An amazing feature of this result is that it predicts an upper limit on the number N_f of different quark flavors. In order that the gluon term be larger than the quark-antiquark term, the value of N_f must be less than or equal to 16. Since the known number of quark flavors is only five or six, it will be some time before this prediction is tested.

Assuming that the number of flavors N_f is less than 16, we see that the vacuum energy from virtual gluons is dominant because of their spin and the larger coupling of gluons. In equation (IX-8), the spin-1 contribution from gluons is twice that of the Dirac spin-$\frac{1}{2}$ field. The coupling of gluons to gluons is also three times larger than the coupling of the virtual quarks to gluons.

The dielectric constant for the QCD vacuum is obtained from Eqs. (IX-5) through (IX-7) and is given by

$$\epsilon = \left[1 + \frac{(33 - 2N_f)}{24\pi^2} g^2 \log\left(\frac{E_2}{E_1}\right)\right]^{-1} \quad . \qquad \text{(IX - 18)}$$

For an ordinary medium, the polarization screens the charge which yields a dielectric constant greater than one. The value of (IX - 18) for QCD vacuum is less than one, which corresponds to antiscreening as we discussed earlier.

In order to apply the result (IX-8) to deep-inelastic scattering, we need to identify E_2 and E_1 with appropriate physical quantities. The energy cutoffs can be related through the uncertainty principle to physical distances $r = 1/E$. To see how the effective color charge changes with r, we define

$$q^2(r) = \frac{q_0^2}{\epsilon(r)} \quad , \qquad \text{(IX - 19)}$$

where q_0 is the "bare charge" in the limit $r \to 0$. We now compare the effective charge values at two different distances r_2 and r_1,

$$\frac{q^2(r_2)}{q^2(r_1)} = \left[1 + \frac{(33 - 2N_f)}{24\pi^2} g^2 \log\left(\frac{E_2}{E_1}\right)\right]^{-1} \quad . \qquad \text{(IX - 20)}$$

For deep-inelastic scattering, the minimum cutoff distance r_2 is determined by the momentum transfer Q. Thus, we can see the effect of increasing Q by holding r_1 fixed and letting $r_2 \to 0$,

$$\frac{q^2(r_2)}{q^2(r_1)} \longrightarrow 0 \quad . \qquad \text{(IX - 21)}$$

We see that the effective color charge, and therefore the strength of the interaction, does indeed decrease as Q increases.

This calculation illustrates another essential difference between QCD and the other gauge theories that we have discussed. In QED, the coupling constant α also varies with distance due to the vacuum polarization by virtual electron-positron pairs. For example, the low energy cutoff for the QED calculation of the Lamb shift is determined by the Bohr radius of the hydrogen atom[22]. Since the Bohr radius depends on the electron mass, we can say that the physical masses

[22] J. D. Bjorken and S. Drell, *Relativistic Quantum Mechanics* (McGraw-Hill, New York, 1964).

for the system provide a natural low energy cutoff. On the other hand, in the Weinberg-Salam theory, vacuum polarization effects arise from the virtual quarks and leptons and the gauge fields just as in QCD. However, the symmetry breaking by the Higgs field generates large masses for the $W^{\pm}$ and Z^0 gauge fields. Thus the vacuum polarization from the virtual $W^{\pm}$ or Z^0 bosons is enormously suppressed. These arguments do not apply to QCD. If we ignore the quarks, which have finite mass, there are still strong interactions between the gluons themselves. The zero mass of the gluons in the QCD vacuum means that there is no corresponding finite mass or other constant to provide an absolute low energy cutoff. We saw in the above calculation for the variation of the effective color charge that it was necessary to use the value of the charge at r_1 in order to provide a reference point to demonstrate the decrease in charge at r_2. Thus an essential difference between QCD and the other gauge theories is that there is no intrinsic mass or energy "scale" in QCD. This behavior shows that QCD actually possesses an unusual type of symmetry known as scale invariance. Roughly speaking, this means that the physical results cannot depend on whether we choose r_1 as a reference point or we "rescale" to some other r value. This scaling property has led to the application of a powerful mathematical approach in QCD which is called the "renormalization group technique". An introduction to this subject can be found in various review articles.[23]

9.5 Discussion

The observation of asymptotic freedom in deep-inelastic scattering provides one of the most compelling arguments that the strong interaction is a color gauge theory. It is evident from the preceding discussion that asymptotic freedom is in fact a unique property of non-Abelian gauge theories[15,16] and therefore may be a valuable guide for further understanding the strong interaction. Unfortunately, it is very difficult to perform either calculations or experimental tests of asymptotic freedom. Since the effective coupling depends only logarithmically on Q, extremely high energies are needed to

[23] A. Peterman, *Phys. Rep.* **53**, 157 (1979).

even detect the change in the coupling.[24]

Asymptotic freedom also provides us with some insight into the properties of hadrons as bound states of quarks. We noted that the spectrum of excited states of charmonium resembled that of positronium. We can now understand the reason why from asymptotic freedom. Evidently, the heavy masses of the charmed quark and antiquark ensure that they are moving relatively slowly in the lowest angular momentum states. The separation between the quarks is then very small and the effective coupling is weak. Under these conditions, the color field is very nearly a Coulomb potential, thus producing the positronium-like spectrum. This description is overly simplistic for charmonium but it may work well for the bound states of new heavier quarks. In constrast, this simple picture clearly cannot apply to the lighter mass hadrons. Since the size of a hadron is inversely related to its mass, most of the known hadrons are relatively large compared to charmonium. The quarks within these hadrons are widely separated and the effective coupling is much stronger. The motion of the quarks must be highly relativistic and the energy states no longer resemble positronium.

Asymptotic freedom may also provide us with a clue as to why free quarks and gluons have not been seen. Let us consider what happens if we attempt to pull a quark out of a proton. As the separation between the quark and the other quarks increases, the effective coupling will also increase and the binding force will become larger. If we assume that the effective coupling can increase indefinitely, we would never be able to extract the quark because it would require an infinite amount of energy. This could explain why free quarks or gluons are not seen in collisions or decays of hadrons. In deep-inelastic scattering, a large momentum Q is transferred to a quark which then moves rapidly away from the remaining quarks. As the separation increase, the effective coupling becomes larger and begins to apply a restraining force. However, if the initial momentum Q is sufficiently large, a new effect will occur. As the

[24] P. Soding and G. Wolf, *Experimental evidence for QCD* (Deutsches Elektronen Synchrotron, Hamburg, DESY 81-013, March 1981).

coupling increases, the energy density in the color field will become large enough to produce quark-antiquark pairs from the vacuum. These quarks have very low energy and will recombine to form mesons and baryons. Similarly, if a high energy gluon is produced in the collision by "bremsstrahlung", the coupling of the gluon with the quarks and the color field will increase in the same way. Thus we see that asymptotic freedom can provide a simple explanation for the fact that high energy collisions produce copious numbers of hadrons but no free colored quarks or gluons. Although this scenario is highly appealing, it is very difficult to prove that the effective coupling increases indefinitely in QCD. The perturbative techniques which work well at small distances cannot be used when the coupling becomes so large that the high-order "corrections" in some series expansion may no longer be smaller than the first order terms. This indicates that certain classes of interesting problems such as the spectrum of the well-known mesons and baryons can never be completely understood with such techniques. The challenge of QCD for the future will be to find new non-perturbative techniques for solving these problems.

CHAPTER X

TOPOLOGY AND GAUGE SYMMETRY

> *The quantization of electricity is one of the most fundamental and striking features of atomic physics, and there seems to be no explanation for it apart from the theory of poles. This provides some grounds for believing in the existence of these poles.*
>
> P. A. M. Dirac, 1948[1]

10.1 Introduction

We have seen in the preceding chapters how the understanding of each of the fundamental forces in terms of gauge theory has led to the discovery of new phenomena such as the Higgs mechanism for symmetry breaking and asymptotic freedom in QCD. Recently an entirely different property of gauge theory has been discovered which may require us to greatly extend our understanding of symmetry. This new property is related to a variety of gauge theory phenomena ranging from the classic Dirac magnetic monopole to the flux quantization in superconductors. What is most significant about this new property of gauge theory is that it originates from a "topological" source rather than from a conventional symmetry.

In this chapter, we present a simple introduction to the basic topological concepts uncovered in gauge theory. Our purpose is to show that it is possible to appreciate at least some of the physical implications of the new topological features of gauge theory without first having to become an expert in topology. We will use the tried and true physicist's approach of starting from a specific familiar physical example, namely the phenomena of flux trapping in a

[1] P. A. M. Dirac, *Phys. Rev.* **74**, 817 (1948).

superconductor, and use it to teach us as much topology as we need.

10.2 Why Flux Trapping?

In many respects, the phenomenon of flux trapping in a superconductor is an ideal pedagogical device for learning about hidden topological concepts in gauge theory. We have already seen in chapter VII that superconductivity itself is one of the simplest real-life examples of a local gauge theory with spontaneous symmetry breaking. As we noted before, the Ginzburg-Landau Lagrangian for a superconductor resembles those of simple models in elementary particle physics. At the same time, one has the advantage that a phenomenon like flux trapping can be described and many of its properties can be calculated with only the use of electromagnetism and elementary quantum mechanics. Thus, no matter how abstract the topology may become, it can always be related to basic physical concepts.

As described in many texts[2], flux trapping occurs in type-II superconductors because the magnetic field induces the flow of a Cooper-pair current. This current forms a vortex which surrounds the magnetic field lines and confines them to a small region where the conductivity is still normal. The flux is quantized according to the condition

$$\text{Flux} = \oint \mathbf{A}\cdot d\mathbf{x} = \frac{2\pi N\hbar c}{q} \quad , \qquad \text{(X - 1)}$$

where q is the Cooper-pair charge and N is an integer. The relation (X-1) is derived from the requirement that the Cooper-pair wavefunction be continuous along any closed path in the superconductor which encircles the flux. The only advanced mathematics required is Stokes' theorem. Thus, in this description, there is no overt evidence of any topological complexities.

How then does one even know that there are any physically

[2] R. P. Feynman, R. B. Leighton and M. Sands, *Feynman Lectures in Physics* (Addison-Wesley, Reading, Mass., 1965).

interesting topological properties to be uncovered in the superconductor? The existence of such properties is suggested by the fact that the choice of the closed path around the trapped flux is completely arbitrary. The quantization condition strongly resembles a contour integral whose value depends only on the residue of a singularity and not on the contour of integration. This resemblance is not entirely accidental. The definition of an analytic function involves a strong connection between the local and global properties of the function over the entire complex plane. A similar kind of global constraint also exists in a system with broken gauge symmetry. We will see in the following discussion how this leads to a relatively simple re-interpretation of flux quantization as a topological condition.

10.3 The Topological Superconductor

Topology is an area of mathematics that has something to do with the global properties of spaces; it tells us why a doughnut is similar to a coffee cup. Gauge theory, on the other hand, is based on the invariance of physical laws under a local internal symmetry. Thus how can a space with global topological properties be uncovered in the superconductor vortex? To find the answer, we must first reformulate the flux quantization problem in gauge theory language. We will then be able to see that the topological properties we are seeking are those of the internal symmetry space associated with the local gauge group.

We saw in chapter VII that the ground state of the superconductor is a "self-coherent" system in the sense that it has a built-in relation between the phase values at different points in space. When magnetic flux is trappped in the superconductor, a potential conflict arises between the intrinsic phase relation of the superconductor and the magnetic field. The magnetic field tries to rotate the local phase of the Cooper-pair wavefunction just as it would the phase of a free particle, but it cannot because the phases are locked together. This conflict is resolved in one of two ways. If the magnetic field is sufficiently strong, the coherence of the Cooper-pairs will be destroyed and the conductivity will return to normal. This is the same type of phase transition that occurs when the temperature is raised above the critical value for a superconductor. The alternative solution is to break the gauge symmetry of the magnetic field and force it to be consistent with the phase relation of the superconductor. This means that the trapped magnetic field, or more precisely the vector potential field **A**, when acting on a free test particle, will produce a phase change in the particle's wavefunction that is the same as the phase relation of the superconductor ground state.

Let us first consider the topology of the internal space without flux trapping since it is easier to visualize. At any fixed position x in the superconductor, the phase of the wavefunction will have a value between 0 and 2π. Continuity of the wavefunction further requires that the phase values of 0 and 2π must coincide. Thus, the internal space at each x is equivalent to a one-dimensional circular loop with the phase being a coordinate on the loop. As we move from x to an adjacent location $x + \mathrm{d}x$, the phase must change continuously in accordance with the intrinsic phase relation of the superconductor. The phase coordinate will therefore trace out a path on the surface of cylinder as shown in Fig. (10.1). Hence, from the point of view of an observer in the internal phase space, the superconductor looks like it has the global topology of a cylindrical space.

When flux is trapped in the superconductor, the topology of the internal space becomes more complicated because the phase relation does not hold in the region where the conductivity is still normal. We can map out the new topology by using the familiar device of a test charge and observing how its phase varies. When

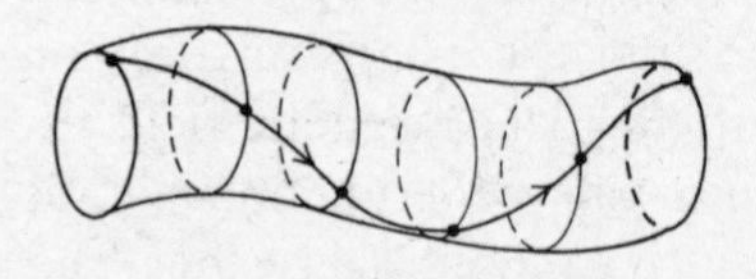

Fig. 10.1 Picture of the internal symmetry space after it has been wrapped around in a circle so that the phase angles of zero and 2π coincide. The phase of a moving particle traces out a path on the surface of a cylinder.

we move the test charge around a closed path encircling the vortex region, we see that it must have the same phase value as it did initially due to the continuity of the superconductor wavefunction. However, the phase of the test charge may have been rotated by the magnetic field through integer multiples of 2π while the charge moved along the path. Thus, in general, we can say that the phase coordinate traces out a path on a surface shaped like a doughnut or torus as shown in Fig. (10.2). It might be argued that this conclusion is obvious because we have only taken the cylindrical space in Fig. (10.1) and wrapped it around a closed path and joined the ends together. However, if there is no trapped flux in the center of the torus, the closed path can be made smaller and smaller until there is no torus. The trapped flux makes a hole in the superconductor so that the path of the test charge cannot be arbitrarily shrunk down to a point. The presence of this hole is therefore essential to the topology of the internal symmetry space. This situation is very similar to the Aharonov-Bohm effect dicussed in chapter II where the electron beams were not allowed to enter the magnetic field inside the solenoid.

Let us now see how the flux quantization condition, Eq. (X-1), can be re-interpreted in terms of the topological properties of the internal space. The vector potential **A** generates a phase rotation given by

$$\delta\theta = \frac{q}{\hbar c}\, \mathbf{A}\cdot d\mathbf{x} \quad . \tag{X - 2}$$

The loop integral in Eq. (X-1) is proportional to the total change in

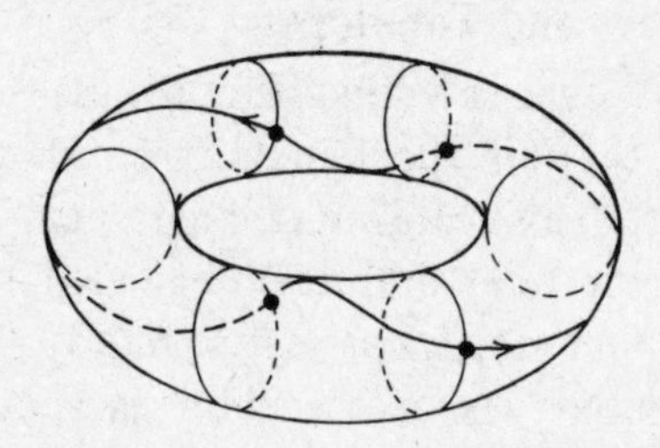

Fig. 10.2 Topological picture of the internal symmetry space for a closed path around the vortex. The cylinder in Fig. (10.1) has been wrapped around the trapped flux and the ends joined together, forming a torus.

in the phase during one complete trip around the vortex. Since the phase can only be rotated by integer multiples of 2π, we see that the flux quantum number N is equal to the number of times that the phase coordinate winds around the torus.

At this point, if we were to communicate our results to a real mathematician, he would tell us that the toroidal internal space we have uncovered is an example of a "multiply-connected" space and that we have also just re-discovered a topological quantity known as the "winding number".[3] What is surprising about this result is not the complicated topology uncovered in the superconductor but rather the fact that a physically measurable quantum number N can be re-interpreted as a topological property of a geometrical symmetry space. It is interesting to compare this with general relativity which shows us that the classical gravitational force can be geometrized. Gauge theory seems to be telling us that certain types of quantum effects can also be interpreted as geometrical, albeit in an unusual type of space.

10.4 The Canonical Vortex

Up to this point, our discussion has relied almost entirely on very intuitive geometrical arguments. In this section, we will compare our geometrical approach with the more formal canonical Lagrangian description of superconductivity. Our purpose is to see what sort of arguments are needed to uncover the topological properties within a Lagrangian formalism.

10.4.1 Internal Space and Topology

In chapter VII, we saw that a convenient Lagrangian for the superconductor is that of the Ginzburg-Landau-Higgs model. How can we now relate this canonical Lagrangian description to our more intuitive geometrical and topological picture of the internal space? Clearly the first step is to locate whatever it is that corresponds to an internal symmetry space in the Lagrangian approach. If we do not assume the *a priori* existence of a geometrical internal space

[3] P. Roman, *Some Modern Mathematics for Physicists and Other Outsiders* Vol. 1 (Pergamon, New York, 1975).

within a strictly canonical Lagrangian formalism, we have only the gauge symmetry group itself to work with. Does this mean that we are now confronted with the problem which we avoided earlier, namely, having to learn abstract topology and more group theory as well? Fortunately, the answer is no because we already discussed in chapter II how to define the internal space directly from the symmetry group itself.

Let us briefly recall the case of the familiar three-dimensional spin-rotation group O(3). In quantum mechanics, an arbitrary spin state can be defined by performing a rotation on a fixed initial spin state. The three angles which specify the rotation can be taken as the coordinates of a point in an abstract three-dimensional space. Each point corresponds to a rotation so that the spin states themselves can then be identified with the points in this angular space. Furthermore, since any rotation can be implemented as a continuous sequence of infinitesimal rotations, one can define "paths" between the points and use these paths to determine the topological structure of the space. If one considers a closed path which starts from one point and returns to the same point, it can be seen that there are two distinct classes of such paths, namely those which can be shrunk continuously down to the starting point and those which cannot. Examples of the two classes of paths are shown in Fig. 10.3[4]. The existence of these two classes is related to the familiar double-valued representations of

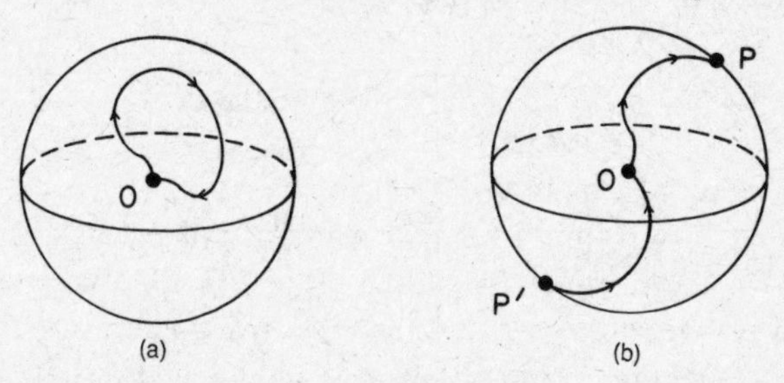

Fig. 10.3 Examples of the two distinct classes of closed paths in the angular space of the rotation group O(3). (a) A path that can be shrunk back down to the point O. (b) The points P and P′ on opposite ends of a diameter represent the same rotation. Thus the path cannot be shrunk to a point.

[4] K. Gottfried, *Quantum Mechanics* Vol. 1 (Benjamin, Reading, Mass., 1966).

half-integer spin. Thus the rotation group has an angular space which is said to be "doubly connected" from a topological point of view.

For the U(1) gauge group, the angular internal space is just the space defined by the local phase values of the superconductor wavefunction as we discussed earlier. This space is just the ring of phases shown in Fig. (10.4). The distinct closed paths in this space are those which wind around the ring a different number of times. Clearly, a closed path which winds twice around the ring cannot be continuously deformed or shrunk so that it only winds around once. Thus, we see that there are an infinite number of distinct classes of paths so that the U(1) group internal space must have an "infinitely connected" topology.

It is possible to associate the distinct classes of paths with the degenerate ground states of the Higgs potentials $V(\phi)$ for the superconductor. Each class of paths represents a phase rotation of $2\pi N$ which we can identify with a gauge transformation. Although arbitrary local gauge transformations cannot be performed on a ground state, rotations of $2\pi N$ are allowed. The reason for this is that a $2\pi N$ rotation has the same effect everywhere, i.e. it is global, and therefore it preserves the intrinsic phase relation of the superconductor. Hence, a $2\pi N$ rotation will transform a ground

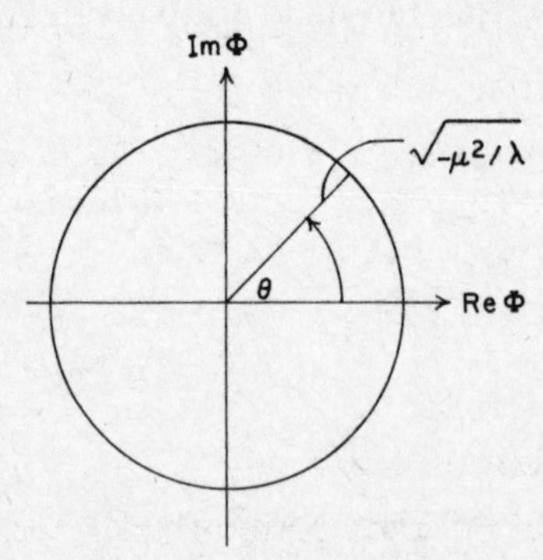

Fig. 10.4 Ring of phase values in the internal angular space of the gauge group U(1). The radius is the magnitude of the ground state wavefunction for the Ginzburg-Landau model of the superconductor.

state into another perfectly legitimate ground state of $V(\phi)$. However, the transformed ground state is not the same as the original ground state because their phases differ by $2\pi N$ which would violate the continuity requirement of the wavefunction. Thus, we can associate each distinct class of paths with a unique ground state, and by doing so, we also associate the connectedness of the internal angular space with the degeneracy of the potential $V(\phi)$.

10.4.2 Where Has the Torus Gone?

The role of the winding number N is evident in the above discussion but there appears to be no sign of the torus in the canonical formalism. In fact, the torus is hidden because the spatial configuration of the vortex has not yet been taken into account. In the geometrical presentation, we saw that the surface of the torus was traced out by the phase of the test charge moving along a closed path around the vortex. No explicit test charge is used in the canonical approach, but the same purpose is served by the axial symmetry of the vortex about the direction of the magnetic field lines. Because of this spatial symmetry, we can perform the following "gendanken" operation: imagine cutting through the torus and collapsing the phase windings together so that they look like a compressed spring. The angular space represented by this collapsed torus is precisely the ring of phases in Fig. (10.4). We note that this gedanken operation is just the reverse of taking the cylinder in Fig. (10.1) and wrapping it around the vortex as we did before. The important point is that there must be two types of closed paths for the torus; one path winds around N times in the internal space, the other path goes once around the vortex in physical space. In the canonical Lagrangian formalism, the physical space path is implicit and only appears in the line integral of Eq. (X-1). Thus, the torus appears to be hidden.

10.4.3 The Degenerate Vortex

The association of each ground state with a distinct class of path, and thus a unique winding number N, leads to an unusual topological picture of the vortex. The vortex can be interpreted as a "transition region" between pairs of different inequivalent ground states. To see how this interpretation arises, we again use a test charge to trace out the ground state phase around the vortex. We start with the test charge very far away from the vortex and align its phase with the ground state wavefunction. We then transport the test charge up to the vortex, move it through the vortex region and see how the phase has changed. From our preceding discussion, we know that the phase of the test charge will be rotated by the magnetic field in the vortex region. Thus, if we move the test charge completely around the vortex and back to its initial position, the phase will have changed by $2\pi N$. Since the test charge is just tracing out the phase of the superconductor ground state, this means that the test charge must have emerged into a different ground state. This leads to a very unusual picture of the vortex as shown in Fig. (10.5). If we imagine that the degenerate ground states are all superimposed on top of each other, then the vortex can be interpreted as a "transition region" between the different layers of ground states. It is amusing that this interpretation is analogous to a Riemann surface with multiple sheets.

A further consequence of the degeneracy is that it provides a simple topological interpretation for the dynamical stability of the vortex. We know that circulating currents actually confine the magnetic field in the vortex. The degenerate ground states prevent the vortex from spreading out and dissipating into the

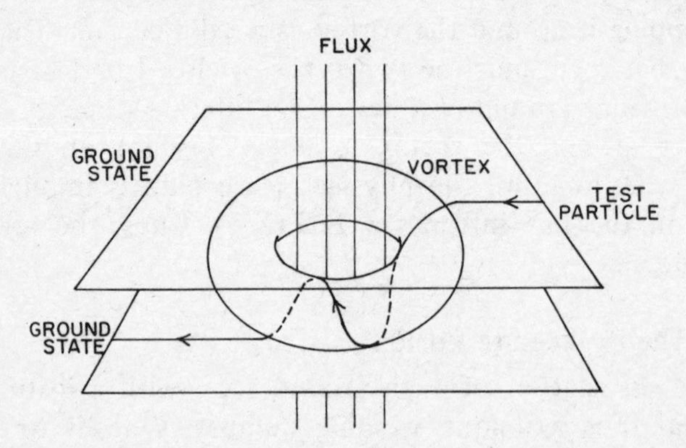

Fig. 10.5 Topological picture of the vortex shown as a transition region between ground states of different winding number.

the surrounding superconductor. The reason for this is that the vortex connects different ground states. If the vortex were to "decay", different ground states would then coincide and violate the continuity requirement for the ground state wavefunctions. Thus, the stability of the vortex can also be interpreted as a topological property.

We see from the preceding discussion that one has to resort to a relatively complicated group theoretical argument within the canonical Lagrangian formalism in order to uncover geometrical or topological properties associated with the superconductor. It is clear, however, that the value of studying the model Lagrangian is that it provides some insights into the relationship between the specific dynamical details of the superconductor and the more general topological concepts.

10.5 How to Add the Flux Number

Let us now use the results of the preceding sections to see how the flux number N can be interpreted as a simple additive quantum number. The question we want to address is the following: given two separate vortices as in Fig. (10.6), with individual flux numbers N and N', can we consider the composite system of two vortices to be equivalent to a single vortex with flux number $N+N'$? To answer in the affirmative, we need to show that the sum $N + N'$ is a valid winding number.

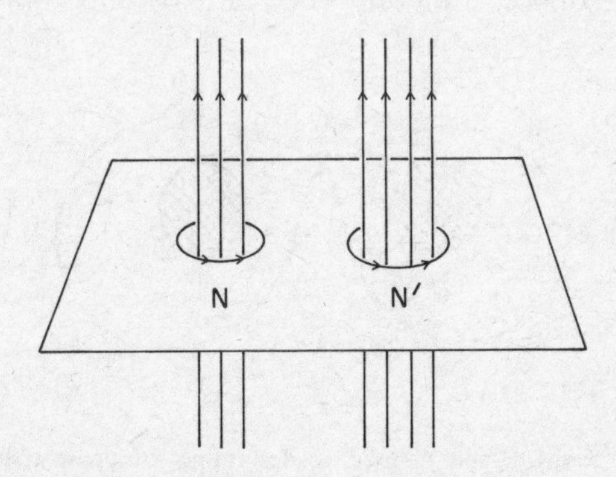

Fig. 10.6 Two separate vortices with different flux numbers N and N'.

We will again use our test charge to determine the properties of the two-vortex system. We want to show that if we move the test charge around each vortex separately, then the net change in phase is the same as that obtained by moving the test charge along a single loop around both vortices. We perform the first part of the operation as shown in Fig. (10.7a). Starting from the point x, the test charge is moved around one vortex along the closed loop C and then around the second vortex along C'. The phase changes along C and C' are $2\pi N$ and $2\pi N'$ respectively. In order to add the phase changes, we must be sure that there is no additional phase change at x when the test charge is switched from C to C'. This question arises because each vortex is surrounded by its own ground states and these ground states must be matched up at x. It is clear that there cannot be a phase difference greater than 2π at x because this would allow a transition between ground states. Phase differences of less than 2π could occur because the ground states are not unique; they actually belong to distinct classes of gauge equivalent ground states. However, equivalent ground states will give the same contribution to the net winding number. We therefore conclude that the flux numbers of the two vortices can be added together to yield a net flux number $N + N'$ for the system.

To see that $N + N'$ is also the flux number for a single path around both vortices, we will show that the loops C and C' can be changed into a single loop. We are allowed to distort the loops as long as the flux is completely encircled by the distorted loops and

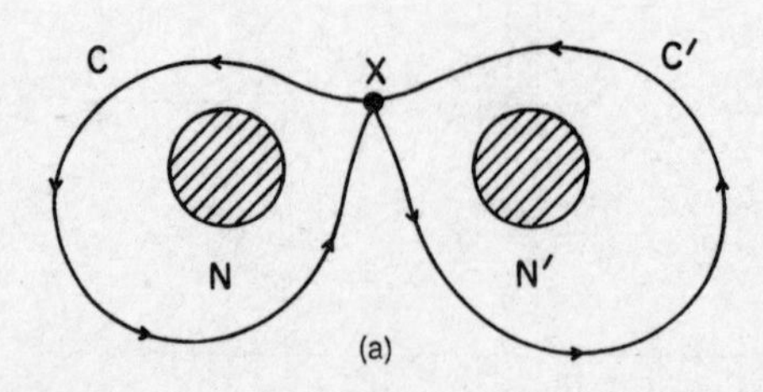

Fig. 10.7a Closed loops C and C' used to determine the phase change around separate vortices with flux numbers N and N' respectively.

the winding number is unchanged. The procedure for altering the loops is illustrated in Fig. (10.7b). The endpoint of C and the starting point of C' are moved from x to a new point y. We are effectively cutting the torus and "patching in" a new piece between x and y. The net flux number remains unchanged because any extra phase change that we have introduced from x to y along C is cancelled by a change from y to x along C'. The path segment between x and y can then be ignored and we are left with a single continuous loop around both vortices. This argument is also reversible since the segment between x and y can be shrunk back down to a point. Thus, we finally conclude that the flux number of the single loop is indeed the same number as the sum of the two loops.

The geometrical path approach for adding flux numbers has the virtue that it leads to a general symmetry structure that is valid for systems other than the vortex. The manipulations we used to merge the loops C and C' into a single loop actually define a new "hidden" symmetry group for the flux number. This unusual group consists of the closed loops themselves with their associated winding numbers plus a definition of the "product" of two loops. A closed loop with winding number N is taken to be the N-th element of the group. The identity or null element is a loop with $N = 0$; it does not encircle any net flux so that it can be shrunk down to a point. The inverse of a loop is another loop which winds in the opposite direction and thus has negative winding number. As we saw above, two loops with winding numbers N and N' can be combined to form

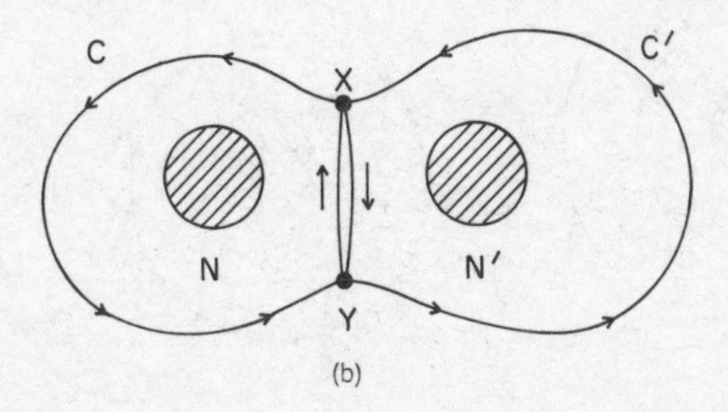

Fig. 10.7b Loops C and C' are converted into a single loop around both vortices by moving endpoints from x to y. Phase changes along the pieces of loop patched in between x and y cancel so that they can be ignored.

a new single loop with winding number $N + N'$. This defines the "product" of two loops. Since the order of the loops is not physically relevant, the product is clearly commutative and the group is Abelian. A product can be formed of loops around separate vortices or around one vortex. For example, the two cases shown in Fig. (10.8) both give a product loop with net flux number equal to zero but with very different physical interpretations. The first case is just a loop which does not circle the vortex while the second case shows a system consisting of a vortex and an anti-vortex with net flux number equal to zero.

The group of closed paths is called the "fundamental group"[5] and is symbolized by

$$\Pi_1 [U(1)] \quad ,$$

where the subscript refers to the one-dimensional closed path in physical space around the vortex. The U(1) gauge group is shown explicitly in the brackets to indicate that the closed loops, which are the group elements, are defined in the U(1) angular internal space. Since each element of the group is labelled by a unique integer winding number N, the group also said to be equivalent to the integers themselves, which form an additive Abelian group (denoted by Z); hence, one can write symbolically

$$\Pi_1 [U(1)] \simeq Z \quad . \qquad (X-3)$$

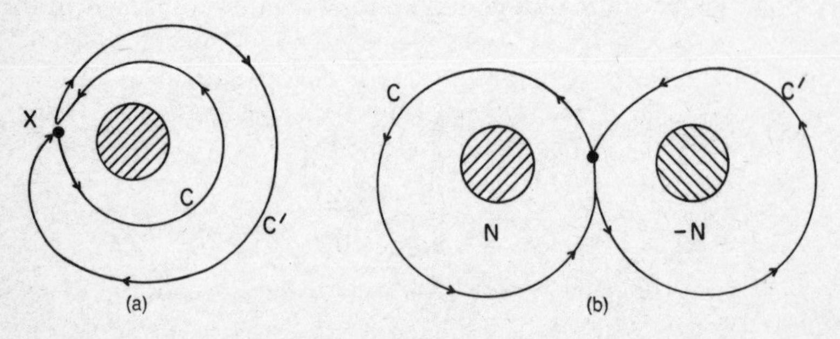

Fig. 10.8 Examples of the product of two loops giving net flux number zero. (a) Product that does not circle the vortex. (b) Product that circles vortex-antivortex system with net flux zero.

[5] G. H. Thomas, *Introductory Lectures on Fiber Bundles and Topology, for Physicists* (Argonne National Laboratory, ANL-HEP-PR-78-23, May 1978).

This equivalence relation neatly summarizes the "hidden" symmetry which underlies the flux quantum number. It also graphically demonstrates just how well the symmetry is hidden from view. A group like Π_1 is rarely encountered in other areas of physics because it is not directly related to any coordinate transformation like most familiar symmetry groups. In fact, Π_1 could be regarded as uniquely characteristic of gauge theory because it arises out of the relationship between sets of paths in the two entirely different types of spaces which are married together by gauge theory.

10.6 The Dirac Magnetic Monopole

One of the most interesting manifestations of the topological ideas we have discussed is the recent resurrection of the classic Dirac magnetic monopole. More than fifty years ago, Dirac[6] attempted to find a fundamental quantum mechanical principle that would explain the quantization of electric charge. Instead, he discovered a surprising relation between the electric charge and the magnetic monopole.

By arguing in analogy with electric charge, the monopole is assumed to be a point-like source of a magnetic "Coulomb field"

$$\mathbf{B} = -4\pi g \boldsymbol{\nabla}\left(\frac{1}{r}\right) \quad , \tag{X-4}$$

where g is the strength or "magnetic charge". By using an electron as a test charge, the magnetic flux can be measured from the change in phase of the electron wavefunction as it is moved around the monopole. Let us calculate the phase change along a closed path C on a spherical surface surrounding the monopole as shown in Fig. (10.9). The net phase change is given by

$$\delta\theta = \frac{e}{\hbar c}\oint_C \mathbf{A}\cdot\mathbf{dx} = 2\pi N \quad . \tag{X-5}$$

In order for the electron wavefunction to be continuous or single-valued, the net phase change must be an integer multiple of 2π. By applying Stokes' theorem, we then see that the net flux Φ from

[6] P. A. M. Dirac, *Proc. R. Soc. London* A**133**, 60 (1931).

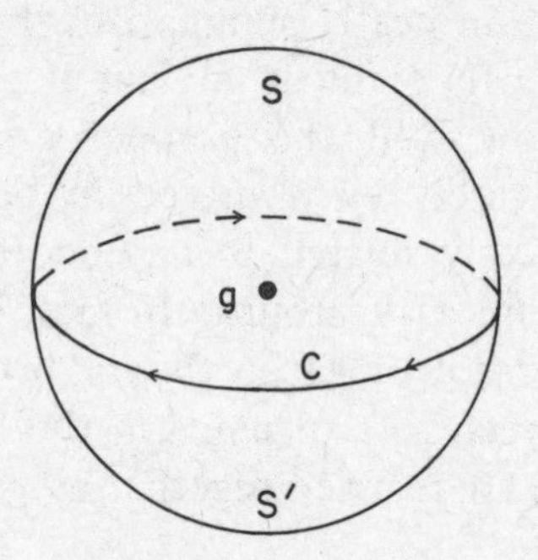

Fig. 10.9 Spherical surface of integration around a Dirac magnetic monopole of magnetic charge g.

the magnetic Coulomb potential must be quantized according to

$$\Phi = 2\pi N \frac{\hbar c}{e} \quad . \tag{X - 6}$$

This relation has the same form as the flux quantization for the superconducting vortex. The electric and magnetic charges are thus related by

$$eg = N \frac{\hbar c}{2} \quad , \tag{X - 7}$$

which is Dirac's famous quantization condition. This equation is remarkable because it states that if even one monopole exists in the universe, then the quantization of electric charge is explained.

The derivation of Dirac's formula is not quite as simple as it seems because there is an essential complication which must be considered in the calculation of the monopole flux. The closed loop C can be taken as the boundary of either one of the two hemispheres S and S' in Fig. (10.9). Thus, we see that

$$\oint_C \mathbf{A}\cdot d\mathbf{x} = \Phi(S) = \Phi(S') \quad . \tag{X - 8}$$

The net flux out of the sphere is the difference of $\Phi(S)$ and $\Phi(S')$ which gives zero and clearly makes no sense. Dirac resolved this contradiction by concluding that there must be some kind of singularity in the magnetic field which passes through the spherical surface. In order to calculate the net flux, a hole has to be cut out of the surface of integration to avoid the singularity. The net flux out of the remainder of the sphere would then be non-zero, thus resolving the contradiction and giving the Dirac quantization condition.

Wu and Yang[7] recently showed that the quantization condition could be derived without explicitly invoking singularities if the magnetic field was re-interpreted as a multiply connected field. The vector potential A around the monopole is considered to be multivalued so that different potential A and A' are used to calculate the flux through S and S' respectively. Along the closed path C, the potentials are related by a gauge transformation

$$\mathbf{A}' = \mathbf{A} - \mathrm{i}\left(\frac{\hbar c}{e}\right)U^{-1}\boldsymbol{\nabla}U \quad , \tag{X - 9}$$

where the transformation U has the form

$$U = e^{\mathrm{i}\lambda(x)} \quad . \tag{X - 10}$$

In order for U to be a single-valued along C, the phase factor $\lambda(x)$ is required to satisfy the condition

$$\oint_C \boldsymbol{\nabla}\lambda \cdot \mathbf{dx} = 2\pi N \quad . \tag{X - 11}$$

This equation takes the place of the continuity condition on the electron's wavefunction. Using Eq. (X - 9) in Stokes' theorem then gives the phase change

$$\begin{aligned} \delta\theta &= \frac{e}{\hbar c}\oint_C [\mathbf{A}' - \mathbf{A}]\cdot \mathbf{dx} \\ &= \oint \boldsymbol{\nabla}\lambda \cdot \mathbf{dx} = 2\pi N \\ &= \frac{e}{\hbar c}[\Phi(S) - \Phi(S')] \quad , \end{aligned} \tag{X - 12}$$

[7]T. T. Wu and C. N. Yang, *Phys. Rev.* D**12**, 3845 (1975); *Nucl. Phys.* B**107**, 365 (1976).

which leads once again to Dirac's quantization condition.

The Wu-Yang derivation appears to give the impression that there are no singularities in the field. However, it has been argued by Barut[8] using different techniques that the singularities must be real. If this is so, how can a singularity be equivalent to a multiply-connected field? We can see the answer to this question by using a simple analogy from the theory of complex variables. Let us consider the case of a function such as log z which has a branch cut singularity. It is well known[9] that if we replace the complex z-plane by a Riemann surface with multiple sheets, then log z can be treated like a single-valued function without an explicit branch cut. A similar situation occurs in the Wu-Yang derivation. The Riemann surface is the analogue of the multiply-connected internal phase space and the gauge transformation, Eq. (X - 9), tell us how the potential changes from one "sheet" to the next. Thus we see that the interpretations of Dirac and Wu-Yang can be perfectly compatible.

The properties of the monopole and the vortex provide an interesting constrast. The flux quantization conditions are essentially identical, yet the Dirac monopole is supposed to be a fundamental particle like the electron while the vortex is a complicated dynamical system. In addition, the Dirac monopole is not related to any gauge symmetry breaking. There are no degenerate ground states associated with the monopole; thus, the multi-valuedness of the field cannot be blamed on the degeneracy of the ground states and the monopole cannot be interpreted as a "transition region" like the vortex. In fact, one might say that the monopole resembles a vortex-like system in which all of the underlying physical mechanisms have been hidden. Thus, the topological properties of the monopole appear more "abstract" than those of the vortex.

[8] A. Barut, *Sov. J. Part. Nucl.* **10**, 209 (1979).

[9] R. V. Churchill, *Complex Variables and Applications*, 273 (McGraw-Hill, New York, 1960).

10.7 Discussion

We have seen that an amazingly rich topological structure can be uncovered from a systematic study of the well known phenomenon of flux trapping. By using the familiar pedagogical device of a test charge and simple geometrical arguments, the flux quantum number N is shown to be associated with a new type "hidden" symmetry. The new symmetry is not based on the usual type of transformation laws but rather is topological in origin.

The two examples of the vortex and the Dirac magnetic monopole demonstrate that topological quantum numbers can arise in very different types of physical systems. We saw that the flux quantization equations were essentially identical, yet the underlying physics could hardly be more different. The vortex is an intricate dynamical system while the monopole is supposed to be an elementary particle like the electron. This contrast between the vortex and the monopole clearly shows that the topological quantum number is independent of the details of the particular system and is based on very general principles.

The discovery of physically relevant topological properties in gauge theories has stimulated many new investigations. By examining gauge theory models with more complex non-Abelian gauge groups, 't Hooft[10] and others[11] have found new monopole-like solutions with interesting properties in a variety of gauge models. It has also been suggested[12] that the kind of topological stability seen in the vortex may provide some insight into how quarks are confined inside hadrons.

[10] G. 't Hooft, *Nucl. Phys.* B79, 276 (1974). See also A. M. Polyakov, *JETP Lett.* **20**, 194 (1974).
[11] A. Jaffe and C. Taubes, *Vortices and Monopoles* (Birkhauser, Boston, 1980).
[12] S. Mandelstam, *Proc. J. H. Weiss Memorial Symp. on Strong Int.* (Univ. Wash., Seattle, 1978).

APPENDIX

SOME KEY GROUP THEORY TERMS

It has been rumoured that the "group pest" is gradually being cut out of quantum physics.
H. Weyl, 1930[1]

A.1 Continuous Groups

The groups which are most useful in gauge theory are known as continuous groups. These groups possess properties which are very different from the more familiar discrete groups such as the set of permutations of n objects or the various crystal symmetry groups. Continuous groups contain an infinite number of elements. They also have unique properties such as being differentiable or analytic like ordinary functions. In this appendix, we present a brief introduction to some key concepts in the theory of continuous groups. Since a comprehensive treatment of this subject would fill at least a volume, we will only present the main ideas. For further details, standard references such as the texts by Gilmore[2] or Hamermesh[3] should be consulted.

A simple example of a continuous group is the set of all complex phase factors of a wavefunction in quantum mechanics. A phase factor can be written

$$U(\theta) = e^{i\theta} \quad . \tag{A-1}$$

The product of two phase factors is

$$U(\theta)\,U(\theta') = U(\theta + \theta') \quad , \tag{A-2}$$

[1] H. Weyl, *The Theory of Groups and Quantum Mechanics* (Dover, New York).
[2] R. Gilmore, *Lie Groups, Lie Algebras and Some of their Applications* (Wiley, New York, 1974).
[3] M. Hamermesh, *Group Theory* (Addison-Wesley, Reading, Mass., 1959).

and the inverse of a phase factor is given by

$$U^{-1}(\theta) = U(-\theta) \quad . \tag{A - 3}$$

These phase factors form a group which is called U(1), for one-dimensional unitary group. The elements of the group are represented by the "one-dimensional" functions (A - 1) which are unitary. Each element of U(1) is characterized by a unique angle θ, which is a continuous parameter, hence the name continuous group. The parameter θ can take an infinite number of values between 0 and 2π so that there are an infinite number of group elements.

The group U(1) is also differentiable. The differential $\mathrm{d}U(\theta)$ can be calculated by changing θ by an infinitesimal amount $\mathrm{d}\theta$, and taking the difference

$$\begin{aligned} \mathrm{d}U &= U(\theta + \mathrm{d}\theta) - U(\theta) \\ &= \mathrm{e}^{\mathrm{i}\theta}(1 + \mathrm{i}\mathrm{d}\theta) - \mathrm{e}^{\mathrm{i}\theta} \\ &= \mathrm{i}\mathrm{e}^{\mathrm{i}\theta}\mathrm{d}\theta = \mathrm{i}U\mathrm{d}\theta \quad . \end{aligned} \tag{A - 4}$$

Clearly, the derivative of $U(\theta)$ is also an element of the U(1) group.

Other examples of continuous groups are familiar in physics. The group of rotations in three-dimensional space, O(3), is a continuous group. The parameters of O(3) are the three rotation angles. The group of Lorentz transformations is another example.

A.2 Compact Lie Groups

The most interesting continuous groups, from the point of view of gauge theory, are called Lie groups, after their inventor Sophus Lie (1842-99). The distinguishing characteristic of a Lie group is that the parameters of a product must be analytic functions of the parameters of each factor in the product. Thus, for any group product

$$U(\gamma) = U(\alpha)U(\beta) \quad , \tag{A - 5}$$

the parameter γ must have the general form

$$\gamma = f(\alpha, \beta) \quad , \qquad \text{(A - 6)}$$

where $f(\alpha, \beta)$ is an analytic function of α and β. This requirement is trivial for the U(1) group which has only a single additive parameter. For more complex groups, the analytic property is necessary to guarantee that the group is differentiable so that an infinitesimal group element like $dU(\theta)$ in (A - 4) can always be defined.

A "compact" Lie group is one for which the parameters are only allowed to range over a closed interval. The group U(1) is compact because the angle θ is defined over the interval $[0, 2\pi]$. The group O(3) also is compact. The property of compactness is important because it guarantees that the group is unitary (or has a unitary representation). The Lorentz group is an example of a noncompact group. The "boosts" or transformations from one inertial frame to another are represented by non-unitary matrices. The boost parameters η, called "rapidities", are given by

$$\eta = \tanh^{-1} \frac{v}{c} \quad , \qquad \text{(A - 7)}$$

where v is the relative velocity. The parameter η is clearly not restricted to a closed interval.

The only groups used in gauge theory (so far) are compact Lie groups. The restriction to compact groups is due to the fact that internal quantum numbers like isotopic spin all appear to be associated with compact symmetry groups.

A.3 Orthogonal and Unitary Group

A.3.1 Orthogonal Groups

The group O(n) is the group of rotations in an n-dimensional Euclidean space. The elements of O(n) are represented by $n \times n$ real orthogonal matrices. These matrices have $n(n - 1)/2$ independent parameters.

A familiar example of O(n) from classical and quantum mechanics is the three-dimensional rotation group O(3). The group O(3) leaves the distance squared $x^2 + y^2 + z^2$ invariant. A convenient set of parameters for the O(3) group is the set of Euler angles α, β and γ. These angles are familiar from classical mechanics in the treatment of rigid body rotation[4]. A general rotation $R(\alpha, \beta, \gamma)$ about some arbitrary axis can be written as a sequence of three rotations, one for each Euler angle,

$$R(\alpha, \beta, \gamma) = R(\alpha, 0, 0)\, R(0, \beta, 0)\, R(0, 0, \gamma) \quad .$$

The rotation $R(0, \beta, 0)$ and $R(0, 0, \gamma)$ are rotations about the y and z-axes respectively, which are represented by the usual matrices

$$R_y(0, \beta, 0) = \begin{pmatrix} \cos\beta & 0 & -\sin\beta \\ 0 & 1 & 0 \\ \sin\beta & 0 & \cos\beta \end{pmatrix} , \qquad \text{(A - 8a)}$$

$$R_z(0, 0, \gamma) = \begin{pmatrix} \cos\gamma & \sin\gamma & 0 \\ -\sin\gamma & \cos\gamma & 0 \\ 0 & 0 & 1 \end{pmatrix} . \qquad \text{(A - 8b)}$$

A.3.2 Special Unitary Groups

Unitary transformations are well known in quantum mechanics because they leave the modulus squared of a complex wavefunction invariant. The elements of the unitary group U(n) are represented by $n \times n$ unitary matrices. These matrices have determinant equal to ± 1. The elements of U(n) with determinant equal to $+1$ only define the special unitary or unmodular group SU(n). The elements of SU(n) have $n^2 - 1$ independent parameters.

[4]Goldstein, *Classical Mechanics* (Addison-Wesley, Reading, Mass., 1959).

The interesting SU(n) groups encountered in gauge theory are the SU(2) group of isotopic spin and the SU(3) group associated with color. The elements of the SU(2) group can be determined from the linear transformations of a complex 2-dimensional vector (u, v),

$$\begin{pmatrix} u' \\ v' \end{pmatrix} = \begin{pmatrix} a & b \\ c & d \end{pmatrix} \begin{pmatrix} u \\ v \end{pmatrix} . \tag{A - 9}$$

We require that "probability" $|u|^2 + |v|^2$ be left invariant and that the determinant $ad - bc = 1$. This leads to the matrices

$$\begin{pmatrix} a & b \\ -b^* & a^* \end{pmatrix} , \tag{A - 10}$$

which represent the elements of SU(2).

A.3.3 Homomorphism of O(3) and SU(2)

The matrices representing O(3) and SU(2) reveal an important relation between the two groups. The elements of SU(2) can also be associated with rotations in three-dimensional space. By defining the new coordinates (x, y, z):

$$x = \tfrac{1}{2}(u^2 - v^2) \quad ,$$

$$y = \tfrac{1}{2i}(u^2 + v^2) \quad ,$$

$$z = uv \quad , \tag{A - 11}$$

it can be shown[3] that SU(2) transformations leave the squared distance $x^2 + y^2 + z^2$ invariant. Thus a real three-dimensional rotation can be associated with any element of SU(2).

In order to relate the Euler angles in (A - 8) to the parameters of SU(2), one can choose $a = \exp(i\alpha/2)$ and $b = 0$ which defines a rotation $R(\alpha, 0, 0)$ about the z-axis. The choice $a = \cos\beta/2$ and $b = \sin\beta/2$ gives a rotation $R(0, \beta, 0)$ about the y-axis. The general $R(\alpha, \beta, \gamma)$ can then be associated with an SU(2) matrix

$$\begin{pmatrix} \cos\frac{\beta}{2}\, e^{i(\alpha+\gamma)/2} & \sin\frac{\beta}{2}\, e^{-i(\alpha-\gamma)/2} \\ -\sin\frac{\beta}{2}\, e^{i(\alpha-\gamma)/2} & \cos\frac{\beta}{2}\, e^{-i(\alpha+\gamma)/2} \end{pmatrix} . \qquad \text{(A - 12)}$$

The matrix (A -12) defines the relation between the Euler angles of O(3) and the complex parameters of SU(2). However, the matrix does not give a unique one-to-one relation. To see this, let us set $\gamma = 0$ and choose $\beta = 0$ or 2π. In three dimensional space, the values $\beta = 0$ and 2π give the same direction. For $\beta = 0$, we let α go to zero and obtain the unit matrix

$$\begin{pmatrix} 1 & 0 \\ 0 & 1 \end{pmatrix} \qquad \text{(A - 13)}$$

which corresponds to zero rotation. However, for $\beta = 2\pi$ and $\alpha = 0$, we obtain instead

$$\begin{pmatrix} -1 & 0 \\ 0 & -1 \end{pmatrix} . \qquad \text{(A - 14)}$$

Thus, we see that the case of zero rotation in three-dimensional space corresponds to two distinct SU(2) elements depending on the value of β. The matrix (A - 12) therefore defines a relation known as a "homomorphism" or "many-to-one mapping" between O(3) and SU(2), where "many" is two in this case.

One well-known consequence of the SU(2) homomorphism is that the SU(2) matrix representations allow half-integer values of angular momentum. The eigenstates allowed by O(3) have only integer values of angular momentum. Thus SU(2) extends the rotation group to both integer and half-integer angular moments. SU(2) is therefore said to be the "covering group" of O(3).

A.4 Group Generators

A.4.1 Rotation Group

In gauge theory, one of the most useful tools from group theory is the group "generator". A familiar example of a group is the angular momentum operator associated with the rotation group O(3). In quantum mechanics, a rotation of a wavefunction ψ about a direction given by the unit vector $\hat{\mathbf{n}}$ is written

$$R(\theta) = e^{-i\theta\hat{\mathbf{n}}\cdot\mathbf{J}} \quad , \qquad \text{(A - 15)}$$

where $\mathbf{J}$ is an angular momentum operator. For an infinitesimal angle $d\theta$, the rotation can be written to first order as

$$R(d\theta) = 1 - id\theta\hat{\mathbf{n}}\cdot\mathbf{J} \quad . \qquad \text{(A - 16)}$$

Let us consider two successive rotations first about the x-axis and then about the z-axis. The combined rotations can be written

$$(1 - id\theta J_3)(1 - id\phi J_1)$$

$$\equiv 1 - id\theta J_3 - id\phi J_1 - d\theta d\phi J_3 J_1 \quad , \qquad \text{(A - 17)}$$

where only terms up to order $d\theta d\phi$ are retained. Now let us reverse the order of the rotations which gives

$$1 - id\phi J_1 - id\theta J_3 - d\theta d\phi J_1 J_3 \quad . \qquad \text{(A - 18)}$$

The difference between the two orderings of the rotations is given by the commutator of J_1 and J_3,

$$d\theta d\phi [J_1, J_3] \quad . \qquad \text{(A - 19)}$$

We know that the angular momentum operators obey the commutation relation

$$[J_i, J_k] = i\epsilon_{ikm} J_m \quad . \tag{A - 20}$$

Thus the difference (A - 19) is not zero, which shows that different orderings of successive rotations produce different results. The commutation relation (A - 20) indicates that the rotation group O(3) belongs to the general class of Lie groups which are called non-commutative or "non-Abelian". The name "Abelian" refers to groups whose elements all commute with each other. The group of rotations O(2) in two dimensions and the unitary group U(1) are Abelian groups.

A.4.2 Lie Group Generators

The preceding discussion of the rotation group can be easily generalized for an arbitrary Lie group G by replacing the angular momentum operators J_k with a set of new operators F_k known as group generators. The operators F_k "generate" the element of G in the same way that the J_k generate rotations,

$$U = \exp(-i\alpha_k F_k) \quad . \tag{A - 21}$$

In general, the number of generators equals the number of independent parameters in the group G. Thus there are $n(n - 1)/2$ generators for the orthogonal group $O(n)$ and $n^2 - 1$ generators for $SU(n)$. The generators F_k of the orthogonal and unitary groups are represented by hermitian or self-adjoint matrices.

The commutation relation satisfied by the generators F_k is written

$$[F_i, F_j] = i c_{ijk} F_k \quad . \tag{A - 22}$$

This commutation relation uniquely determines the generators F_k and therefore the structure of the group G. Thus the c_{ijk} are known as the "structure constants". The structure constants for an Abelian group are all equal to zero. We see from the commutation relation (A - 20) that the values of the structure constants for the O(3) group are equal to either ±1 or 0.

The generators F_k have a mathematical structure that is very different from the group itself. The group elements are multiplied together, while the generators are added like the sum $\alpha_k F_k$ in the exponent of (A - 21). The F_k actually form the basis of a linear vector space which is known as a "Lie algebra". This space of generators has both a vector and a scalar product. The commutation relation of the generators defines a form of vector multiplication, which is called a Lie product or more commonly, a cross product. A familiar example of the scalar product of generators is the square of the total angular momentum

$$J^2 = (J_1)^2 + (J_2)^2 + (J_3)^2 \quad . \tag{A - 23}$$

This square is a special case of a scalar product known as a "Casimir invariant" which commutes with all of the generators and is therefore invariant under all group transformations. The Casimir invariants have the useful property that their eigenvalues are the conserved quantum numbers associated with the symmetry group. The number of invariants depends on the particular group. The O(3) and SU(2) groups have only one Casimir invariant while SU(3) has two invariants. For a compact group, the Casimir invariant can always be written as a sum of squares of generators as in (A - 23).

A.5 The SU(3) Group

In this section, we summarize some useful properties of the SU(3) group. As we noted above, SU(3) has eight independent generators. A convenient representation for the generators is the following set of 3×3 hermitian matrices:

$$F_1 = \frac{1}{2}\begin{pmatrix} 0 & 1 & 0 \\ 1 & 0 & 0 \\ 0 & 0 & 0 \end{pmatrix} \qquad F_2 = \frac{1}{2}\begin{pmatrix} 0 & -i & 0 \\ i & 0 & 0 \\ 0 & 0 & 0 \end{pmatrix}$$

$$F_3 = \frac{1}{2}\begin{pmatrix} 1 & 0 & 0 \\ 0 & -1 & 0 \\ 0 & 0 & 0 \end{pmatrix} \qquad F_4 = \frac{1}{2}\begin{pmatrix} 0 & 0 & 1 \\ 0 & 0 & 0 \\ 1 & 0 & 0 \end{pmatrix}$$

$$F_5 = \frac{1}{2}\begin{pmatrix} 0 & 0 & -i \\ 0 & 0 & 0 \\ i & 0 & 0 \end{pmatrix} \qquad F_6 = \frac{1}{2}\begin{pmatrix} 0 & 0 & 0 \\ 0 & 0 & 1 \\ 0 & 1 & 0 \end{pmatrix}$$

$$F_7 = \frac{1}{2}\begin{pmatrix} 0 & 0 & 0 \\ 0 & 0 & -i \\ 0 & i & 0 \end{pmatrix} \qquad F_8 = \frac{1}{2\sqrt{3}}\begin{pmatrix} 1 & 0 & 0 \\ 0 & 1 & 0 \\ 0 & 0 & -2 \end{pmatrix} . \tag{A-24}$$

The generators F_3 and F_8 commute with each other. They are related in the SU(3) quark model to the isotopic spin operator I_3 and the hypercharge Y,

$$I_3 = F_3, \qquad Y = \frac{2}{\sqrt{3}} F_8 \quad . \tag{A-25}$$

The structure constants c_{ijk} for $i < j < k$ are listed in Table 1 below. The other values can be obtained by using the fact that c_{ijk} is antisymmetric under the interchange of any pair of indices.

Table 1: SU(3) Structure Constants

ijk	c_{ijk}	ijk	c_{ijk}
123	1	257	$\frac{1}{2}$
147	$\frac{1}{2}$	345	$\frac{1}{2}$
156	$-\frac{1}{2}$	367	$-\frac{1}{2}$
246	$\frac{1}{2}$	458	$\frac{3}{2}$
		678	$\frac{3}{2}$

INDEX